Managing Projects in Research and Development

Managing Project Delivery
Maintaining Control and Achieving Success

Dr Trish Melton
Dr Peter Iles-Smith

AMSTERDAM · BOSTON · HEIDELBERG · LONDON · NEW YORK · OXFORD · PARIS
SAN DIEGO · SAN FRANCISCO · SINGAPORE · SYDNEY · TOKYO

Butterworth-Heinemann is an imprint of Elsevier

Butterworth-Heinemann is an imprint of Elsevier
Linacre House, Jordan Hill, Oxford OX2 8DP, UK
30 Corporate Drive, Suite 400, Burlington, MA 01803, USA

First edition 2009

British Library Cataloguing in Publication Data
A catalogue record for this book is available from the British Library

Library of Congress Cataloging-in-Publication Data
A catalog record for this book is available from the Library of Congress

ISBN: 978-0-7506-8515-3

For information on all Butterworth-Heinemann publications
visit our web site at www.elsevierdirect.com

Typeset by Charon Tec Ltd., A Macmillan Company
(www.macmillansolutions.com)

Printed and bound in Great Britain

09 10   10 9 8 7 6 5 4 3 2 1

About the authors

Trish Melton is a project and business change professional who has worked on engineering and non-engineering projects worldwide throughout her career. She works predominantly in the chemicals, pharmaceuticals and healthcare industries.

She is a Chartered Chemical Engineer and a Fellow of the Institution of Chemical Engineers (IChemE), where she was the founder Chair of the IChemE Project Management Subject Group formed in 1998. She is a part of the Membership Committee which reviews all applications for corporate membership of the institution, and in 2005 she was elected to the Council (Board of Trustees).

She is an active member of the International Society for Pharmaceutical Engineering (ISPE) where she served on the working group in charge of updating ISPE's *Active Pharmaceutical Ingredient (API) Baseline® Guide*. She was the founder Chair of the Project Management Community of Practice (PMCOP), formed in 2005. She has presented on various subjects at ISPE conferences including project management, quality risk management and lean manufacturing and has also supported ISPE as the conference leader for project management and pharmaceutical engineering conferences. She is also the developer and lead trainer for ISPE's project management training course. In 2006, the UK Affiliate recognized Trish's achievements when she was awarded their Special Member Recognition Award. In 2007, she was also honoured for 'Outstanding Leadership and Service', related to her work on the launch and development of the PMCOP.

Trish is the Managing Director of MIME Solutions Ltd, an engineering and management consultancy providing project management, business change management, business improvement, regulatory and GMP consulting primarily for pharmaceutical, chemical and healthcare clients.

Within her business, Trish is focussed on the effective solution of business challenges, and these inevitably revolve around some form of project: whether a capital project, an organizational change programme, a business improvement using Lean Six Sigma or an interim business solution. Trish uses project management on a daily basis to support the identification of issues for clients and implementation of appropriate, sustainable solutions.

Good project management equals good business management, and Trish continues to research and adapt best practice project management in a bid to develop, innovate and offer a more agile approach.

Peter Iles-Smith is a project management and engineering professional who has worked in the field of manufacturing automation and IT projects in the oil and gas, chemical and pharmaceutical industries for vendors and users. The last 5 years have been spent developing strategies, projects and technologies to improve pharmaceutical manufacturing.

Peter is a Chartered Chemical Engineer and a Fellow of the Institution of Chemical Engineers (IChemE). He was a founder member of the IChemE Project Management Subject Group and is the current Chair. He is a former Chair of the IChemE Process Management and Control Subject Group and has served on Council twice. He is a founder committee member of the ISPE PMCOP and former Education Chair of WBF (World Batch Forum), the forum for manufacturing professionals.

In his current position, he is responsible for the automation projects strategy within the technical and engineering functions of a major pharmaceutical manufacturer.

About the Project Management Essentials series

The Project Management Essentials series comprises four titles written by experts in their field and developed as practical guidelines, suitable as both university textbooks and refreshers/additional learning for practicing project managers:

➤ *Project Management Toolkit: The Basics for Project Success.*
➤ *Project Benefits Management: Linking Projects to the Business.*
➤ *Real Project Planning: Developing a Project Delivery Strategy.*
➤ *Managing Project Delivery: Maintaining Control and Achieving Success.*

The books in the series are supported by an accompanying website: www.icheme.org/projectmanagement, which delivers blank tool templates for the reader to download for personal use.

Foreword

This book has become a reality for a number of reasons:

- As experienced Project Managers, we realized that more and more we were dealing with customers, sponsors and project team members who had no project management experience. The first book in this series, *Project Management Toolkit*, was a direct response to that. However we have found that project delivery is a particular area where expertise is needed.
- As founders of the IChemE Project Management Subject Group (PMSG) and then more recently a part of the Continuous Professional Development (CPD) and Publications Sub-groups, it was also evident that there wasn't a full series of books which would support the further development of Project Managers.

Managing Project Delivery: Maintaining Control and Achieving Success is intended to be a more in-depth look at the third value-added stage in a project and builds on from Chapter 5 of *Project Management Toolkit* (Melton, 2007).

The other books in the project management series are outlined earlier (page vii).

Although this book is primarily written from the perspective of engineering projects within the process industries, the authors' experiences both outside of this industry and within different types of projects have been used extensively.

The tools, methodologies and examples are specific enough to support engineering managers delivering projects within the process industries; yet generic enough to support the R&D manager in launching a new product, the business manager in transforming a business area, the IT manager in delivering a new computer system or the Lean Six Sigma practitioner in delivering step change business improvements. The breadth of the short and full case studies demonstrates the generic use of these delivery methodologies over a wide range of industries and project types.

Project delivery is achieved by people and impacts people. All projects start life as separate entities to the normal business environment. However, if they are planned and delivered with little consideration of, or consultation with, those involved or affected, the impact is felt long after the 'hard' side has been forgotten. This book demonstrates the importance of integrating the hard and soft elements of project delivery, ensuring that 'no project is an island'.

Dr Trish Melton and Dr Peter Iles-Smith

Acknowledgements

In writing a book which attempts to go into greater detail and to share a greater level of expertise than previously (*Project Management Toolkit*), you need to effectively develop that expertise – gain peer review of that expertise and then share and test it. We therefore want to acknowledge a number of people against these specifics:

For supporting the development of project delivery expertise over many years:

➤ All past colleagues and clients.

For supporting the peer review of this collated project delivery expertise:

➤ Bill Wilson, Astrazeneca.

For sharing and testing this collated project delivery expertise on real 'live' projects:

➤ All current clients, in particular Paul Burke, Astrazeneca.
➤ Associates of MIME Solutions Ltd such as Victoria Bate, Andrew Roberts and others. In particular we want to thank Andrew for his insight and contribution of some of the unique project challenges when improving a business using Lean Six Sigma techniques.

Finally we would like to acknowledge the support of our families, in particular our partners Andrew and Heather and the special contribution of Katie.

Authors Note: Although all the case studies presented in this book are based on real experiences they have been suitably altered so as to maintain complete confidentiality.

How to use this book

When you pick up this book we hope that before you delve into the content you'll start by glancing here.

The structure of the book is based on the concept that every project goes through three types of delivery phases – these are described in Chapter 1 and then each phase becomes the subject of its own chapter (Chapters 3–5).

Chapter 1 is a general introduction to the concept of controlled project delivery. This can be read at any time to refresh you on some basic concepts which are applied within the core chapters. This chapter also provides the link between the *Project Management Toolkit* (Melton, 2007) and this book, which is a more in-depth look at the third value-added stage in a project.

Chapter 2 reflects on the 'thing' being delivered and considers the similarities and differences in delivering a stand-alone project, a programme or a portfolio of projects.

Chapters 3–5 are the 'core chapters' made up of the following generic sections:

- Introduction of detailed delivery concepts.
- Presentation of specific methodologies and how they support effective delivery.
- Introduction of project delivery tools and associated tool templates.
- Demonstration and/or further amplification of chapter concepts, methodologies and tool use through the use of short case studies.
- Presentation of troubleshooting notes and a summary of handy hints.

Each core chapter can 'stand-alone' so the reader can dip into any delivery phase.

Chapter 6 is a collation of short case studies which demonstrate a lack of delivery management success and analyse 'why?'

Chapters 7 and 8 contain a series of fuller case study projects, and in effect is the culmination of the use of all the areas of expertise introduced in the previous chapters. These aim to show the breadth of project delivery issues that may arise and how these have been successfully dealt with. Within these case studies various formats for project delivery progress reporting are presented, based on the needs of the specific project or programme.

The blank project delivery tool templates can be accessed via either of the following two websites: http://books.elsevier.com/companions or www.icheme.org/projectmanagement. The actual format of the template cannot be changed but the tool can be used electronically by the reader to fill in the project data as required.

And remember . . .

➤ As a Project Manager your goal is to deliver the project as planned, taking into account the current environment – both internally within the project and externally within the business.

➤ Plans are only value add if they are used during delivery.

➤ Plans change over time. It is better to plan the change than explain the failure.

➤ Project delivery is about change – both internally within the project and externally within the business. The role of the Project Manager is to facilitate that change to deliver the benefit.

Contents

4 Set-up plan delivery

5 Control plan delivery

1 Introduction

This book develops the project delivery concepts originally outlined in *Project Management Toolkit* (Melton, 2007).

Following the approval of the Project Delivery Plan (PDP), the Project Manager and Project Team are ready to commence project delivery. At this stage, it is crucial that the plan is followed in order to manage uncertainty and increase the probability of success.

The role of a Project Manager at this stage in a project is to be in control – of both the 'hard' and 'soft' side of a project. This means management of team and stakeholder relationships, the link to the business environment, as well as the controlled delivery of scope within agreed cost, time, quality, quantity and functionality targets.

The project lifecycle

As outlined in *Project Management Toolkit* (Melton, 2007), a project goes through four distinct 'value-added' stages from its start point to its end point (Figure 1-1). Each stage has its own start and end point, and each has a specific target to achieve. Effectively each stage can be considered a 'project' within a project.

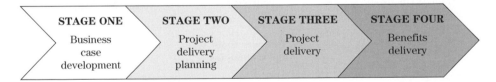

STAGE ONE	STAGE TWO	STAGE THREE	STAGE FOUR
Business case development	Project delivery planning	Project delivery	Benefits delivery

Figure 1-1 The four 'value-added' project stages

Stage One: Business case development

The project start point is usually an idea within the business, for example an identified need, a change to the status quo or a business requirement for survival. At this stage, the project management processes should be challenging whether this is the 'right' project to be progressing.

Stage Two: Project delivery planning

This stage is all about planning and the project management processes are used to determine how to deliver the project 'right'.

Stage Three: Project delivery

Effective delivery is all about the control and management of uncertainty. This stage is therefore focused on the controlled delivery – to deliver the project 'right'.

Stage Four: Benefits delivery

The final stage involves integrating the project into the business – allowing the project to become a part of the normal business process, business as usual (BAU).

This book is concerned with Stage Three: project delivery, where the start point is typically approval to deliver a project (an approved PDP) and the end point is a successfully delivered project.

Aims

The aim of this book is to introduce the need for controlled project delivery to an audience of Project Managers who have had both good and not so good experiences when delivering their projects. It provides the reader with education, tools and the confidence to deilver projects so that the chances of success are increased.

Figure 1-2 shows an input-process-output (IPO) diagram for this book.

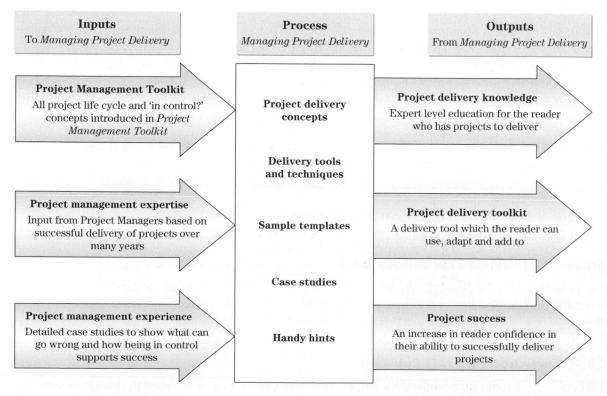

Figure 1-2 The IPO for this book

Figure 1-2 represents the process by which the aims are to be achieved:

⮞ *Inputs* – lists the inputs to the development of this book.
⮞ *Process* – summarizes the contents of this book.
⮞ *Outputs* – lists the outputs from this book from the perspective of the reader.

Although there are many 'basic' delivery tools and techniques available, the aim of this book is to introduce methodologies and principles to support the delivery of more complex projects and programmes. Initially though, it will reinforce the project delivery tools and concepts introduced in *Project Management Toolkit* (Melton, 2007):

⮞ Asking 'in control?' is fundamental to the third value-added stage in a project and one that is asked throughout the project delivery process.
⮞ Delivery of a project involves more than controlling the cost, scope and time, the 'softer elements' must be considered. A good Project Manager must effectively integrate all these elements and then manage and forecast so that the project remains in control.
⮞ Controlled delivery is all about increasing the chances of success in an uncertain world.
⮞ The process of project delivery links Stage Two with Stage Four; it aims to enable the realization of the business benefits, which have been 'promised' within the approved business case.

The book will continue to develop generic tools and techniques which can be applied within any type of organization and any type of project. This will be demonstrated in Chapters 7 and 8 through different case study projects. In addition, Chapter 6 demonstrates different facets of project failure in terms of ineffective delivery.

What is project delivery?

Ask most people this question and they will clearly say 'it's the doing of the project'. And whilst this is true, there is much more complexity to successful project delivery.

It is clear from many project experiences that project delivery is traditionally seen as the management of three project constraints: cost, scope and time. Assuming a realistic baseline has been set, the role of the traditional Project Manager has been to restrict change and deliver the baseline. This model of operation earned many successful Project Managers the title of 'control freak' due to strict change control mechanisms they put in place. However, being in control is much more than the control of change. A Project Manager needs to ensure that he can answer the following questions:

⮞ **Why?** – Is the project still going to meet the needs of the business case?
⮞ **What?** – Is the project still able to deliver its scope-based goals?
⮞ **What if?** – Are there scenarios within which the project cannot deliver?
⮞ **When?** – Is the project still able to deliver its time-based goals?
⮞ **How?** – Is the PDP being followed in terms of the defined delivery methodologies?
⮞ **How much?** – Is the project still able to deliver its resource-based goals (funding, assets and people)?
⮞ **Who?** – Are the Project Team and all stakeholders delivering on their commitments?

Therefore, the question 'in control?' is much more complex than it might at first appear. It is in fact an assessment of the uncertainty within which the project is operating and its continuing ability to respond to the reason why it was needed in the first place.

Short case study

A business change project within a manufacturing organization was approaching 50% completion. The project concerned the automation of some aspects of a manufacturing plant and other changes in the way the operations team worked. The general view was that the project was progressing according to plan and was well managed and controlled. This was backed up by the following project metrics:

- Schedule adherence – 100%, all milestones achieved to date.
- Cost adherence – 95%, slight under spend in terms of overall spend committed versus plan.
- Deliverables – 90%, minor delay to completion due to final approvals bottleneck.
- Change log – 15 changes logged, 4 rejected, 9 approved and 2 under review.
- Issues log – no project issues logged.

However, during a routine progress check some additional questions were asked (Table 1-1).

Table 1-1 Case study progress update

Project:	Plant X Operational Improvement	Project Manager:	John Smith	Date:	Month 3
Progress check		**Response**			
1. Why? – is the project still going to meet the needs of the business case?		The approved business case involved a capital spend to deliver a capacity improvement, which required no changes to the plant manning levels. The current solution appears to be capable of delivering this.			
2. What? – is the project still able to deliver its scope-based goals?		The scope is divided into three areas: - Automation – scope on track. - Engineering modifications – some delays on change approvals. - Standard Operating Procedures (SOP) – some delays on SOP approvals. Root cause of delays appears to be the Operations Team (management and team members).			
3. What if? – are there scenarios within which the project cannot deliver?		If the issues with the Operations Team are not resolved, the project may be unable to deliver the scope which is required to enable the capacity improvements. This appears to be about the team's resistance to changes in their ways of working. This is an issue which should have been logged.			
4. When? – is the project still able to deliver its time-based goals?		The approvals which are missing were not on the project critical path, however they were 'near critical' and within a week the critical path will be impacted.			
5. How? – is the PDP being followed in terms of the defined delivery methodologies?		The contract plan is working, with both engineering and automation suppliers working well together. The Project Team appears to be following the PDP although the risk log and project report only appear to focus on 'hard' metrics/issues.			
6. How much? – is the project still able to deliver its resource-based goals (funding, assets and people)?		The method of tracking cost versus planned cash flow shows that the progress is slowing down – an indicator of future schedule issues.			
7. Who? – are the Project Team and all stakeholders delivering on their commitments?		The Project Team has not truly engaged the Operations Team or the Operations Director (sponsor) and concerns about the achievement and sustainability of the business benefits should be raised. Unless the Operations Team buy into the changes and are committed to operating the plant in the 'new way' then the capacity improvements will not achieve target levels.			

In many respects, the project was 'in control'; however, the level of overall uncertainty had changed and the probability of success had decreased. The majority of the changes to the risk profile were linked to the 'softer' side of project delivery.

- Not engaging the end-user yet allowing this team's approval to delay progress.
- Not engaging the sponsor to support the above and the ultimate sustainability of change.
- The lack of teamwork between the Project Team, main suppliers and Operations Team (the end-user).

Based on this slightly different review, the Project Manager was able to turnaround the current situation by having a team day. This day brought together all the main stakeholders and allowed all to gain a better understanding of why? what? what if? etc. As a result, the Operations Team were able to challenge the solution, understand their role in the project and deliver what was needed. In the end, the project was successfully completed and the business benefits sustainably delivered. In this case, this meant the successful delivery of capacity improvements through the Operations Team following the agreed new ways of working.

Delivery of 'soft' and 'hard' elements

What the previous case study demonstrates is that all aspects of a project need to be delivered if a project is to be completely successful: both 'soft' and 'hard' elements.

- Soft – generally these refer to people, behaviours, relationships and intangible parts of the project or business case.
- Hard – generally these refer to the more tangible elements in a project: scope, cost, time, project deliverables and financial benefits.

The effective integrated delivery of both elements is what makes a project a success. Project management is the management of uncertainty and the Project Manager's goal is to progressively increase the certainty of outcome as delivery progresses. In managing the delivery stage appropriately, the Project Manager will hopefully deliver a predictable and successful outcome. This management of uncertainty is crucial for an organization as a whole as part of good business management involves forecasting the outcome of all activities.

The target for Stage Three is to complete all aspects of project delivery in such a way which:

- Delivers the project scope to meet the business needs.
- Ensures that the expected outcome is delivered for the business.
- Ensures that the business is ready to receive the project and to deliver the benefits.
- Delivers the appropriate options which increase the probability of success.
- Demonstrates appropriate control.

Why we need controlled project delivery

To understand why a project should be delivered in control, we have to understand why good project management practices are needed: to prevent chaos at any stage in a project's life cycle. Project chaos is often described as 'utter confusion' and the symptoms we typically see are:

- Projects delivered late or outside of their agreed budgets.
- Projects which do not deliver to agreed quality, quantity and functionality criterion.
- Projects which do not meet the intended business needs.

It is easy to react to the variety of symptoms, but such a reaction can lead to further issues. In order to develop sustainable and robust project management practices, the root cause of any symptom needs to be found and resolved (Figure 1-3).

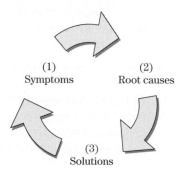

Figure 1-3 Symptoms and root causes

There are many techniques which can be used to identify root causes, but the one used here is 'five whys':

- Ask 'why?' a maximum of five times.
- With each 'why?' the cause becomes more specific and therefore actionable.
- Usually the first and second 'why?' will generate further symptoms.
- Usually the third or fourth 'why?' will generate the cause of the specific project issue.
- Typically the fifth 'why?' will generate the root cause which requires resolution at the organizational level.

Within this chapter (page 8), the 'five whys' technique has been used to identify project management practices within Stage Three of a project that need to be used to deliver project success: the delivery of sustainable business benefits.

Chaos in Stage Three – project delivery

Examples of typical symptoms of project chaos in Stage Three are shown in Figure 1-4. The majority of these are seen during either project or benefits delivery, leading to the conclusion that chaotic or poor project management impacts delivery effectiveness.

It is therefore useful to review key activities within the delivery stage which define when a project is 'in control?', how it is to remain that way and what can happen when these are not robustly performed. Typically these are:

- Scope management through the management of deliverables and activities in terms of appropriate quantity, quality and functionality.
- Cost management through the management of cost expenditure per planned item, cash flow profile and cost risk spend/run down.
- Schedule management through the management of the critical and near critical path, critical milestones and schedule buffer use/run down.
- Risk management through delivery and review of appropriate mitigation and contingency actions.
- Contract management through the delivery and review of appropriate contract plans, ensuring that organizations work well together to successfully deliver the project.

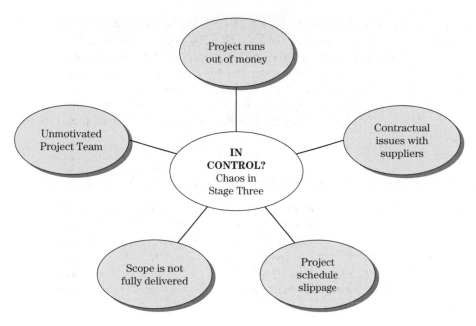

Figure 1-4 Symptoms of project chaos in Stage Three 'in control?'

- Stakeholder management through appropriate engagement and communication activities – both within and external to the Project Team.
- Health checks of all aspects of project delivery in order to continually assess whether the project is in control, to forecast its future status and whether it is likely to be successfully completed.

One example of an issue seen during project delivery is when a project simply runs out of money. This is demonstrated by the following short case study.

Short case study

A project to upgrade and expand various areas of a chemicals manufacturing site was suspended when only 50% of the installation work was complete. The Project Steering Team highlighted the following reasons for this suspension:

- Forecasts showed an unacceptable level of over spend (greater than 15%).
- A general lack of confidence in cost and schedule forecasts to completion.
- A concern that the Project Team was 'out of control' particularly with regard to supplier management.

The Steering Team then invited an external consultant to conduct a formal review and recommend any changes to general project practices. He was asked to offer a view on what should be done with the suspended project in order to maximize business benefit realization through minimal additional investment above the approved funding.

Initially, a series of 'five whys' analyses (Figure 1-5) were performed in order to determine the root cause of the situation. These were conducted with an extended team made up of Project Team members, end-users and occasionally key suppliers and subcontractors (particularly where there was

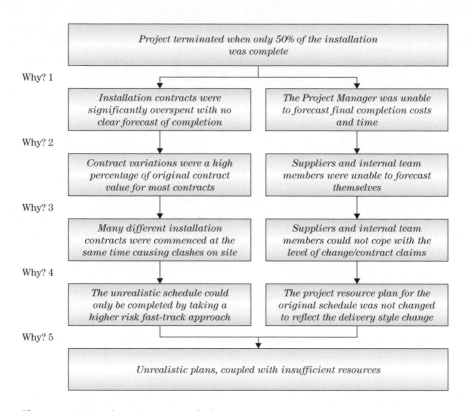

Figure 1-5 Example root cause analysis

a 'partnership' approach). In this case, the root cause for the lack of control in project delivery was a combination of two areas of poor project management:

1. *Unrealistic plans* – The contract plan and associated cost and schedule plans were not integrated (a change in one was not reflected in the other). This lead to inappropriate baselines from which to track progress. The team recognized that there was no point in trying to achieve them and so stopped tracking very early on in the project.

2. *Insufficient resources* – The level of internal and supplier resources was always based on an original plan to deliver the project over a much longer period with minimal crossover of installation subcontracts. The eventual fast-track approach required more management time for all concerned: more contract management, team management and general contract co-ordination. This was never allowed for.

Throughout these sessions, it became clear to the consultant that there was, generally, a culture of low trust between engineers and senior management, particularly when developing business cases for capital projects. The engineers felt they had to overestimate benefits and cut investment costs, otherwise the project would not get approved. The view was 'the management always want more for less' and so they behaved in a way that made this true. However, the Project Team made no attempt to manage to these apparently unrealistic goals. The consultant therefore recommended the following changes:

➤ Project sponsors must become accountable for the delivery of the business case and ensure that it is realistic.
➤ Project Managers and sponsors must work more closely together to deliver the project and maintain control.

- A specific set of project metrics must be used to demonstrate control, and the validity of the associated forecasts must be assured.
- Project Managers and Project Team members should have specific project targets within their own personal development plans.
- A set of standard project control tools should be developed/used on all future projects.

The consultant worked with the Project Team to identify the area of scope within the suspended project which, when completed, would deliver the greatest benefits. He then developed the following for that area of scope:

- A realistic cost, scope and schedule baseline from which to track progress.
- A project review process which would highlight any critical deviations.

The remainder of the project was cancelled and works were only completed if there was a safety or product quality issue. By the end of the project about two-thirds of the original scope was delivered within the original budget but 6 months later than planned. In the year following project completion, the capacity sustainably increased by 50% versus the target of 60%.

Generic delivery issues

Table 1-2 shows the results of a few more generic 'five whys' analyses which equally point to an issue in Stage Three as the root cause for lack of project success. The common theme through all the solutions for the root causes is the concept of controlled project delivery management – the theme of this book.

Table 1-2 Example root cause analyses – when a project is not 'in control?'

Typical symptoms	Example root causes	Example solutions
Project 'out of control' as soon as delivery starts	Lack of contingency plans to cope with project issues; Lack of risk management	Ongoing assessment and tracking of risks (and *risk profile*) so that *contingency actions* can be progressed
Project runs out of money	Lack of appropriate progress tracking	A clearly delivered *progress tracking methodology* which highlights resource issues (cost, people, assets) and risks to completion
Contractual issues with suppliers	Lack of appropriate contract management and contract progress tracking	A clearly delivered *contract management methodology* with output tracking metrics
Project schedule slippage	Lack of appropriate progress tracking	A clearly delivered *progress tracking methodology* which highlights schedule issues and risks to completion
Scope is not fully delivered (quantity, quality and/or functionality)	Lack of appropriate progress tracking	A clearly delivered *progress tracking methodology* which highlights scope issues and risks to completion
Unmotivated Project Team	Lack of appropriate people (individual and team) management	A clearly delivered *team management methodology* which highlights team issues, progress and risks to completion
Project treated like it's an 'island'	Lack of communication between the Project Team and the business	Management of *stakeholders* and their level of engagement so that the project remains robustly linked to the business
Scope delivered does not enable the required business benefits	Lack of change control processes to protect the delivery of value	Management of value so that *scope and business change control* are maintained and tracked through use of appropriate metrics

These solutions either explicitly or implicitly suggest that the symptom would be eliminated through the delivery of appropriate progress tracking methodologies. These are needed to ensure that:

- The project is being delivered to meet a set of specific business requirements.
- The project delivery is as effective as possible: considering both 'soft' and 'hard' issues.
- The project remains 'out of chaos' and therefore 'in control'.

Control in this context is the ability to predict the outcome rather than the ability to do everything you planned to do. So to answer to the general question 'why do we need controlled project delivery?' the succinct answer is 'to ensure that it has the highest potential of achieving its specified outcome; the highest chance of project success'.

The delivery hierarchy

There is a clear vision of success for Stage Three of a project (Figure 1-6) and an associated path of critical success factors (CSFs), which is analogous to those for project delivery planning (Melton, 2008).

CSF 1 Approved PDP A robust PDP providing business, set-up and control plans appropriate to a project so that the Project Manager has a clear and detailed set of delivery methodologies	CSF 2 Engaged stakeholders The management and engagement of the sponsor, customer, Project Team and user groups are necessary to deliver a successful project	CSF 3 Capable Project Manager A project requires an appropriately experienced Project Manager using proven skills, knowledge and behaviours to deliver a successful outcome
CSF 4 Business plan delivery Management and control of external stakeholders and business case delivery through robust two-way links with the business	CSF 5 Set-up plan delivery Management and control of the internal stakeholders (Project Team) and the delivery of value	CSF 6 Control plan delivery Management and control of uncertainty through progress tracking, forecasting and management of risk

Vision of success
The controlled delivery of a customized PDP (that describes the most appropriate methodology to deliver a specific project) which increases the potential for success through management of uncertainty

Figure 1-6 Project delivery success

CSF 1 – Approved PDP

This CSF is the start point for successful project delivery as it sets the baseline for all subsequent delivery decisions. It would be usual for the PDP to be developed by the Project Manager selected to deliver a project. The PDP should then have been reviewed so that the way a project is to be delivered has been thoroughly challenged. One method to complete this review would be by using the 'How?' Checklist Tool (Appendix 9.2) introduced in *Project Management Toolkit* (Melton, 2007).

CSF 2 – Engaged stakeholders

This CSF recognizes that delivery is a team activity; an extended team activity requiring inputs from various sources. Delivery is an ongoing opportunity to engage with a broad stakeholder base whilst providing people with current information on the project and its potential impact on them and/or their business units. It is equally an opportunity to gain information/feedback from the stakeholder group. Effective two-way communication is an indication of stakeholder engagement.

CSF 3 – Capable Project Manager

This CSF recognizes that success at any stage in a project not only needs good organizational support (infrastructure and processes), a good Project Team and a sound basis for the project but also a capable Project Manager. During Stage Three, the Project Manager is responsible for delivering the approved PDP, communicating with all stakeholders and achieving project delivery success.

CSFs 4, 5 and 6 – Business, set-up and control plan delivery

In terms of project delivery, CSFs 4, 5 and 6 can be considered together within a delivery hierarchy (Figure 1-7). This hierarchy proposes three levels of delivery, that are critical to project success, each linked to a separate CSF.

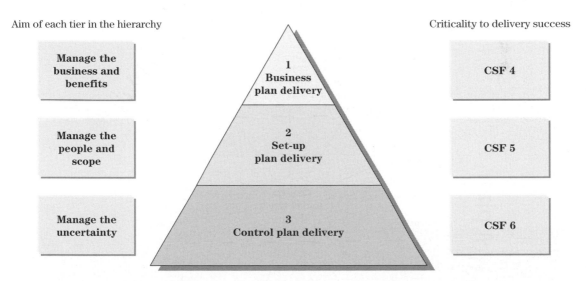

Figure 1-7 The delivery hierarchy

Earlier in this chapter, the reasons why 'in control' delivery is needed were highlighted, and these are all related to the prevention of project chaos and therefore project failure. Delivery needs to be controlled and uncertainty managed because the Project Manager and the business want to:

- Deliver the business case – *Business plan delivery* (*CSF 4*).
- Control what should be delivered, who should be involved and in what way – *Set-up plan delivery* (*CSF 5*).

- Control when delivery should occur within predetermined cost, quality, quantity and functionality criterion – *Control plan delivery* (*CSF 6*).

The Delivery Hierarchy (Figure 1-7) is therefore at the heart of the concept of successful project delivery and encompasses these three levels so that the certainty of a successful outcome is increased and project chaos eliminated:

- **Business plan delivery** – effective management of the link between the project and the business.
- **Set-up plan delivery** – robust project administration.
- **Control plan delivery** – controlled project delivery.

A suite of project management tools and techniques are required to deliver all three plans in parallel.

The project health check

The start of a well-controlled project is to have a PDP which aims to maximize success by planning to manage all areas of uncertainty (Figure 1-8). Each area of uncertainty needs to be managed and these 11 planning and delivery themes were originally introduced in *Project Management Toolkit* (Melton, 2007)

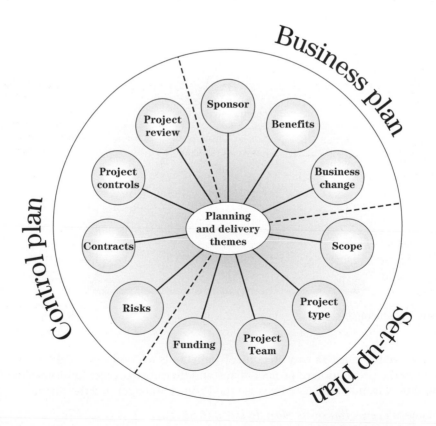

Figure 1-8 Areas of uncertainty – components of a project delivery health check

by use of the 'How?' Checklist (Appendix 9.2), and the basic tools used address them. The 11 themes represent the main components of a PDP and also a project health check.

Although a detailed PDP will have been developed and approved prior to the start of project delivery, it is still necessary to conduct project health checks at key stages in the project, including immediately following project launch. Project health checks are a positive and proactive way to check that the project delivery progress is appropriate. They can also be considered:

» A formal status review between the project sponsor and the Project Manager – checking the status of the link between the project and the business.
» A communication and management tool for the Project Manager and the Project Team – checking the status of the internal 'workings' of the project.
» A communication tool between the Project Manager/Project Team and the external project stakeholders – further integrating the project within the organization in which it sits through frequent and appropriate communications on the health of the project.

The 'In Control?' Checklist (Appendix 9.3) introduced in *Project Management Toolkit* (Melton, 2007) is one method of assessing the health of a project; however, others are also appropriate. In this book, the 11 delivery themes have been divided (by chapter) into the three tiers of delivery (Figures 1-7 and 1-8), and within each tier additional concepts and tools will be introduced (Table 1-3) to support proactive delivery and project review during delivery.

Table 1-3 Summary of delivery concepts and tools

	Business plan delivery (Chapter 3)	Set-up plan delivery (Chapter 4)	Control plan delivery (Chapter 5)
Concepts	» Business strategy management » Sponsorship management and engagement » Communication management » Stakeholder management and engagement » Benefits management and risk assessment » Business change management and risk assessment » Sustainability management	» Set-up strategy management » Capability management and risk assessment » Team conflict management » Virtual team working » Project road map management » Project funding and finance management » Scope and deliverable management » Value management	» Control strategy management » Risk response management » Tracking and forecasting of risk, cost, change, contracts and schedule » Contract and supplier management » Quality control » Theory of constraints and lean six sigma » Project performance management and Visual Factory
Tools	» Business Plan Review Checklist » Communications Tracker » Benefits Totalizer » Benefits Delivery Fault Tree » Sustainability FMEA » Stakeholder Engagement Tracker » Business Change Mitigation Matrix	» Tracking RACI » Team Issues Tool » Project Team Audit Tool » Scope Tracker	» Project Risk Profile » Project HACCP Evaluation » Project HACCP Hazard Analysis and Control » Issue Action Manager » Contract Tracker » Supplier Performance Evaluation » Supplier Performance Tracker » Project Control Tracker » Project Waste Analysis Tool » Project Diagnosis Fishbone

However, the concept of challenging the progress and completion of a project through using the 'In Control?' Checklist remains valid and a reinforcing concept throughout each of the case studies (Chapters 7 and 8).

The remainder of this book is structured around the concept of the Delivery Hierarchy (Figure 1-7) and the 11 delivery themes (Figure 1-8). It proposes a 'health check' methodology which:

➤ Assesses the effectiveness of the links between the project and the business through the tracking of business benefits and elements of project delivery, which support enabling and sustaining these benefits.

➤ Identifies the project status within the context of scenarios of uncertainty and therefore can forecast a likely outcome.

➤ Focusses equally on the 'hard' and 'soft' elements of project delivery.

As well as introducing concepts and tools, short case studies are used in each chapter to further demonstrate key aspects of business, set-up and control plan delivery.

And remember . . .

➤ There are four valued-added stages in the project lifecycle.

➤ Stage Three is all about the controlled delivery of the project in alignment with the PDP.

➤ Controlled project delivery should mitigate typical causes of project chaos often introduced at Stage Three.

➤ At the end of Stage Three, the project should have been delivered and it should be clear how the project outcome will support the realization of a set of specific business benefits.

➤ The progress and conclusion of Stage Three can always be checked via use of the 'In Control?' Checklist which was introduced in *Project Management Toolkit* (Melton, 2007).

2 Project, programme and portfolio delivery

As described in *Real Project Planning* (Melton, 2008), within project management, the terms project, portfolio and programme are used to describe specific activities:

- *Project*: A bounded piece of work which is non-routine for the organization. It is not a part of business as usual (BAU) but has a defined start and end point (when it is integrated into BAU).
- *Programme*: A set of interdependent projects working together to achieve a defined organizational goal. There is dependency between project outputs/benefits.
- *Portfolio*: A collection of projects using a common resource pool. These resources could be assets, people or funding. There is a dependency between these project resources, which therefore need to be used optimally.

However projects, programmes and portfolios all use similar project management processes in order to deliver to agreed criteria:

- *Activity based criteria* – all need to deliver a specified scope which requires specific activities to be completed.
- *Resource based criteria* – all have limited, rather than limitless, access to organizational resources (people, funds, assets).
- *Benefits based criteria* – all are required to deliver a set of benefits to the business.

What's the difference?

Although there are different definitions in common use for projects, programmes and portfolios, Figure 2-1 presents a simple picture of how the three fit together in one organization.

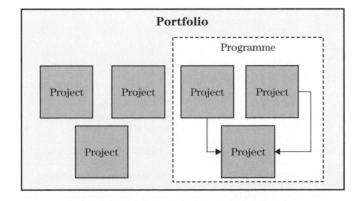

Figure 2-1 How projects, programmes and portfolios interact

A portfolio can be made up of any number of projects or programmes in various stages of delivery, and a programme can comprise a number of interdependent projects. A large organization may have a number of portfolios across departments or project types (Figure 2-2). In this way a hierarchy of portfolios develops within an organization, each portfolio separate from each other and each managing their own projects and programmes.

Table 2-1 summarizes key similarities and differences which impact how delivery should be managed to optimize success.

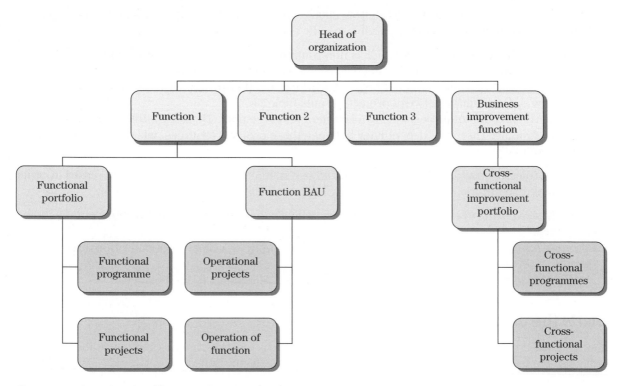

Figure 2-2 Hierarchy of portfolios

Table 2-1 Projects, programmes and portfolios explained

	Project	Programme	Portfolio
Key characteristics	➧ SMART project objectives ➧ SMART benefit metrics	➧ A set of projects which together deliver a set of SMART benefit metrics	➧ A set of a specific types of project within an organization
Managed by	➧ Project Manager	➧ Programme Manager and a team of Project Managers	➧ Portfolio Manager and a team of Project Managers

(*Continued*)

Table 2-1 (Continued)

	Project	Programme	Portfolio
Project selection rationale	• Approved business case • Organizational cost/benefit analysis	• Contributes to programme goal • Prioritization of project activities versus benefit metrics	• Contributes to organizational goal • Availability of required resources • Prioritization of resources versus benefits criteria
Key delivery goals	• Controlled delivery of a specified set of project objectives • Stakeholder engagement of the project activities and benefits outcomes	• Controlled delivery of a specified set of programme objectives • Optimal flow of activities linked to optimal flow of benefits • Stakeholder engagement of the programme activities and benefits outcomes	• Delivery of a balanced portfolio delivering the highest priority organizational benefits • Optimal use of resources
Key delivery challenges	• Managing change and project stakeholders	• Managing dependencies and programme stakeholders	• Managing resource prioritization and organizational stakeholders
Delivery baseline	• Project delivery plan	• Programme delivery plan	• Portfolio delivery plan
Control focus	• Activity management	• Dependency management	• Resource management
Typical measures that demonstrate 'control'	• Progress in line with plan • Low-risk project (or trending to lower risk) • Sponsors and stakeholders engaged • Benefits realized	• High-level dependent events' critical path progressing in line with plan • Sponsors and stakeholders engaged • Low-risk profile across dependencies • Benefits realization profile in line with plan • Project Managers engaged	• Good flow of projects through the portfolio in line with plan • Appropriate balance in the portfolio as linked to the business drivers for that portfolio • Effective and efficient use of resources • Prioritization appropriate • Project Managers engaged
Typical behaviours to support delivery management	• Structured, focussed approach • Ability to build, motivate and manage a Project Team	• Strategic, focussed approach – linked to programme drivers • Ability to engage and influence stakeholders • Ability to build, motivate and manage a Programme Team	• Strategic, focussed approach – linked to portfolio drivers • Ability to engage and influence stakeholders • Ability to build, motivate and manage a team of Project Managers
Delivery process	• Project management	• Programme management	• Portfolio management

Delivering projects, programmes and portfolios

Although much of this book is about the delivery of projects, the majority is also applicable to the delivery of programmes and portfolios. However, one of the main differences is in the nature of the overall process that is followed.

Project management process

Every project goes through four value added stages (Figure 1-1), although there are many different project roadmaps linked to the specific stages and stage gates that a project must progress through. Some of these stages are linked to overall governance of the project process. In this book the project delivery themes are described (page 12) and are then explained in some detail. The project management process is valid for all projects within either a programme or a portfolio.

Programme management process

There would usually be a programme roadmap which describes the interdependencies of the individual projects and the stage gates may be linked to these (Figure 2-3). Each project would be delivered according to an appropriate project roadmap and still go through each of the four generic value added stages. In this way it is clear that the project delivery themes remain valid for the delivery of each of the projects within the programme: project management within a programme is still project management.

The programme management process is then the management of the programme roadmap:

- Understanding which projects are linked to others in the programme and what that link enables.
- Managing the stage gates at project start and end or at defined points within delivery.
- Ensuring that each individual project is managed according to best practice project management so that they are capable of delivering their part of the programme.
- Understanding and tracking the benefits realization profile as the programme progresses (Figure 2-4).

The programme used as an example in Figures 2-3 and 2-4 demonstrates the need for good benefits management:

- The Pareto and dependency charts show that 80% of the required benefits can be delivered by four of the six projects that are currently in the programme (P1 and P3, P2 and P4).
- Project 2 (P2) appears to be a 'quick win' project that may have non-tangible 'softer' benefits by preparing the business area for bigger changes.
- At stage gate 1 the Programme Manager can work with the business to decide if project 6 (P6) should be delivered by approving the delivery of project 5 (P5).

Portfolio management process

Portfolios are usually a collection of projects using a similar project roadmap, for example an engineering project portfolio or a business change portfolio. The process is effectively structured around the management, co-ordination and prioritization of resources so as to maximize the delivery of business benefits for an organization through the optimal mix and sequencing of proposed projects. Portfolios are usually formalized:

- To support more effective programme and project management (as a form of governance).
- To enhance communication between projects, programmes and the business.
- To support decision-making on the best use of resources versus the benefits to be realized (this may be a mix of financial and non-financial benefits).
- To provide transparency on the scale and breadth of projects within an area of an organization.

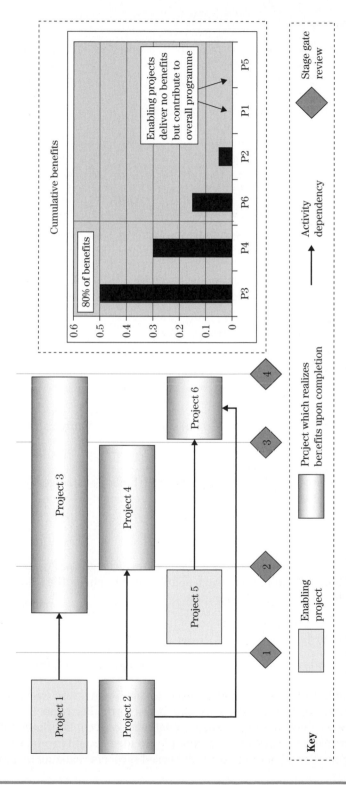

Figure 2-3 Example of a programme roadmap

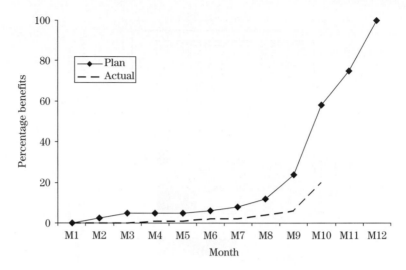

Figure 2-4 Example programme benefits profile

Initially a portfolio needs to be designed and then a steering process put in place. The design is based around a number of parameters.

Portfolio driver

The reason why a set of projects is to be managed as a portfolio is usually called the portfolio driver. Examples of these are:

- Use of common resources (people, assets, funds).
- Delivery of similar project scope (engineering, research, business change, IT).
- Focus on a specific area of the business or a specific set of business benefits.

Portfolio balance

A portfolio usually needs to be balanced so that it fully achieves its goals. For example, Figure 2-5 shows how a business change portfolio has used a spider chart to check the total number of projects delivering changes in each of five areas of operation.

Portfolio progress tracking

There are a number of methods and metrics to track portfolio progress. Figure 2-6 shows one such metric – portfolio speed.

In this example the number of projects in each project stage is reviewed every 6 months to check how quickly projects are moving through the portfolio. It also provides a check on whether the portfolio needs refreshing with new projects in order to use available resources. Bear in mind that there is usually project attrition as the portfolio progresses. For example, 10 ideas could produce 9 business cases which could result in 6 approved projects giving 6 delivered projects realizing benefits for the business.

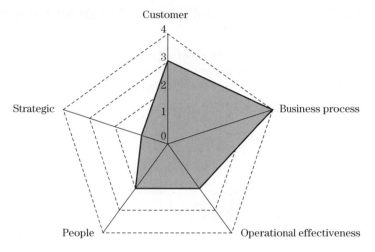

Figure 2-5 Portfolio balance

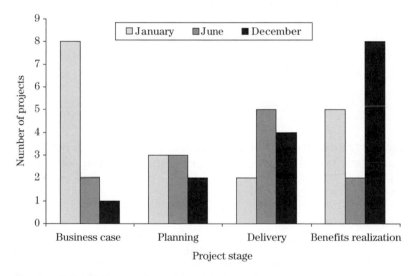

Figure 2-6 Portfolio speed

Portfolio prioritization processes

A portfolio is by definition a defined (potentially constrained) set of resources delivering a collection of projects. As projects enter and progress through the portfolio, prioritization is needed:

- What are the key drivers defining decision points to approve, hold, stop or reject projects?
- Portfolio decisions tend to be made based on project progress, business priorities and resource availability.

Tools and techniques

Project management tools and techniques remain valid for the delivery of projects within either programmes or portfolios; however, different or adapted tools are needed to manage the cumulative impact of projects in either. Within programmes and portfolios there are similar techniques used to assess the cumulative progress, for example:

- ➤ *Benefits charts* – quantitative trending charts showing metrics trending across a number of projects (Figure 2-4). Within a portfolio this tends to be the way to trend financial benefits, for example. Whereas in a programme there may be other non-financial metrics that can still be collated such as cycle time reduction in a manufacturing process.
- ➤ *Risk profile charts* – a method of reviewing the risk status of each project in a portfolio or programme and presenting the cumulative data as a pie chart (Figure 2-7).

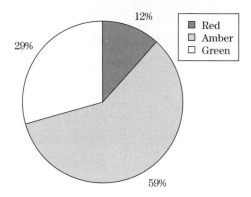

Figure 2-7 Portfolio or programme risk profile chart

Table 2-2 summarizes a typical programme and portfolio management toolkit. As the emphasis on both is to manage a cumulative set of benefits (whether interdependent or not), most tools can be used for either.

Table 2-2 Programme and portfolio toolkit

Tool	Description	Comment
Dependency Map	A PERT chart showing either benefit or project objective links and dependencies	Useful within programmes where a project, task or benefit critical path may need to be defined
Dependency Chart (resources)	A table showing all resource dependencies within a programme, such as when one project requires the same internal or external resources	Typically used within portfolios in order to optimize use of resources
Benefits Mapping (portfolio or programme)	A map showing how benefits relate to each other, to the portfolio or programme goals and to higher organizational goals	A customized Benefits Map would be generated for each programme or portfolio
		Useful in assessing changes, additions or deletion to a programme or portfolio

(Continued)

Table 2-2 (Continued)

Tool	Description	Comment
Benefits Pareto Chart	Linked to benefits scoring. Used to assess if the resources are being appropriately used within the portfolio. For instance, are 80% resources being used to deliver 80% benefits	Useful if the benefits are scored considering both financial and non-financial benefits Within a portfolio it can support project selection and prioritization Within a programme it can support project sequencing
Benefits Specification Table (portfolio or programme)	A table defining how specific benefits criteria can be measured (benefits metrics) and the parts of a programme or a portfolio which should deliver this	Requires a Benefits Map to be generated
Benefits Scoring Tool (portfolio)	A tool to assess relative priority of independent benefits based on the Organizational Benefits Map	Should consider financial and non-financial benefits which support project selection
Benefits Scoring Tool (programme)	A tool to assess relative priority of dependent benefits based on the Programme Benefits Map	Should consider projects which enable other projects even if they deliver no actual benefits themselves
Portfolio or Programme Matrix	A benefits matrix tool as applied to a specific portfolio or programme that identifies which projects in the portfolio/programme support achievement of specific portfolio/programme objectives or benefits criteria/metrics	A tool to demonstrate alignment that can also be used to highlight progress or risks by using traffic lights; can be a visual indicator for the team.
Portfolio Flow Rating	A tool to assess the speed at which projects are moving through a portfolio versus benchmark KPIs	The way to use this is to assess the resource constraints and look at benchmark delivery metrics

Portfolio and programme management processes rely on the effective use of all project and portfolio/programme tools:

- The foundation connecting a project to a portfolio or programme is the suite of benefits identification and management tools.
- The focus of management is the stage gate(s) (and stage gate milestone plans and decisions), benefits score and risk rating.
- Measures/metrics should be based on benchmark KPIs for similar scale/size or type of projects.
- Customized reporting should be used to communicate the goals of the programme or portfolio effectively. Reporting should not just be the collation of how each project is currently progressing but the cumulative impact of the projects' status on the programme or portfolio goals.

Steering projects, programmes and portfolios

Projects, programmes and portfolios need to be appropriately guided if they are to achieve the required goals as expected by senior management within an organization. This process of guiding is typically referred to as 'steering' and can occur at many levels dependent on the specifics of the project, programme or portfolio.

As discussed in later chapters of this book, sponsors and senior stakeholders usually provide steering for individual projects. This is typically the situation for programmes. Where either is particularly large, critical or complex, separate steering groups may be formed. In this case three types of steering mechanisms are usually used based on the progress of the specific project or programme:

- *'Hard' steering* – making stage gates decisions during delivery. For example approving the design before implementation.
- *'Soft' steering* – giving general advice based on information external to the project. For example letting the Project Manager know that an end user isn't happy with a particular aspect of the project or programme.
- *Exception steering* – when exceptional circumstances arise and a steering decision is needed to further progress the project or programme. For example a major risk occurring or a project defect.

Portfolio steering is quite different. A Portfolio Manager needs to work with the business to make key decisions regarding progress of the portfolio and appropriate use of organizational resources. The steering takes two forms:

1. *Resource steering* – the allocation of resources to specific projects. This can be people, assets or funding. For example funding approval and Project Manager allocation. During the delivery of the portfolio specific resources will be allocated to one or more projects and may be moved from one project to another.
2. *Prioritization steering* – the adjustment of the sequence of projects based on business drivers. For example, a project may be put on hold even if it is progressing well because another project has a higher priority and needs the resources. This is common in research situations when a pilot plant 'slot' may be taken from one project and used in another.

Managing projects, programmes and portfolios

At the heart of any process is the person who remains responsible for delivery: the Project, Programme or Portfolio Manager. Simplistically the following comparisons could be made:

- A Project Manager thinks about tasks and his team = tactical thinking.
- A Programme Manager thinks about benefits and his stakeholders = strategic thinking.
- A Portfolio Manager thinks about operational goals and his resources = operational thinking.

The three roles are considered different career paths, requiring different skills and capabilities, and different experiences (Table 2-3).

Table 2-3 Role comparison

Role comparison		
Project Manager	**Programme Manager**	**Portfolio Manager**
➡ Managing tasks being delivered by members of a Project Team	➡ Managing projects being delivered by Project Managers	➡ Managing resources so that they are optimized to deliver organizational goals
➡ Managing micro activity dependencies	➡ Managing micro and macro activity and benefit dependencies	➡ Managing macro activity with resource dependencies

(Continued)

Table 2-3 (Continued)

Role comparison		
Project Manager	**Programme Manager**	**Portfolio Manager**
➧ Focus on delivery of the project CSFs	➧ Focus on delivery of business benefits	➧ Focus on achieving organizational goals in the most optimal way
➧ Managing and influencing the project sponsor	➧ Managing stakeholders who expect the delivery of specific strategic benefits	➧ Influencing stakeholders who own resources being used within the portfolio
➧ Managing cost, scope and time ➧ Managing resources and micro resource dependencies ➧ Managing contracts ➧ Managing interfaces ➧ Delivering stage gate requirements and influencing stage gate decisions	➧ Managing people and politics ➧ Managing culture change ➧ Managing strategic change ➧ Managing stage gate reviews and influencing stage gate decisions	➧ Managing project management capability ➧ Managing the project pipeline ➧ Managing stage gate reviews and delivering stage gate decisions
➧ Specific project management tools skills and experience required	➧ Specific capability required linked to 'softer' areas (leadership, communication)	➧ Specific line management skills and experience required
➧ Tactical level	➧ Strategic level	➧ Operational level

Project, Programme and Portfolio Manager capability

In *Real Project Planning* (Melton, 2008), the concept of the development of project management capability was discussed. This focussed on the knowledge, skills and experiences which a Project Manager needed to develop in order to increase capability. Any Project Manager should be considering his overall career path and associated decisions as a part of developmental planning (Figure 2-8).

Figure 2-8 Generic career path

A typical career path for a Project Manager would be to gradually do larger and more complex projects within a specific project type (Figure 2-9); however, there are many other routes that can be followed:

➧ *Broaden project management experiences* – changing to different project types. For example an engineering Project Manager with a strong engineering background may move to delivering business improvement projects within an engineering environment and from there to generic business change projects.
➧ *Expand into programme management* – Project Managers working within business change projects get much earlier opportunities to manage a programme.
➧ *Move into portfolio management* – Project Managers who want more operational line management experience move into this type of role. It still requires project management capability but is focussed on specific skill areas such as people management and development.

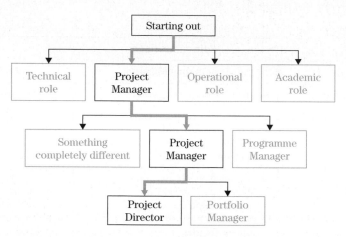

Figure 2-9 A possible project management career route

And finally . . .

➡ Projects, programmes and portfolios have a lot in common, and good project management remains at the core of all three.

➡ Project Managers can move into Programme or Portfolio Manager roles if they choose to develop their project management capability in that direction.

3 Business plan delivery

In the context of project delivery, the business plan is the way that we link the project to the business. It is based on the approved business case which explains 'why' the project is needed in the context of the organization. The robust delivery of the business plan relies on the Project Manager and sponsor working together to ensure that the project being delivered remains linked to the business within which it will deliver benefits. The two concepts which are fundamental to business plan delivery are:

➤ *Stakeholder management* – the continued engagement of all types of stakeholder; from sponsor to customer, end user to team member.
➤ *Business case delivery* – the robust delivery of all aspects of the cost/benefit equation which justified the project in the first place.

In order to manage these concepts the Project Manager needs to continue to act in a consultancy mode as described in *Real Project Planning* (Melton, 2008).

What is a project business plan?

As defined in detail in *Real Project Planning* (Melton, 2008), a project business plan is that part of a Project Delivery Plan (PDP) that is the formal articulation of HOW the project will link effectively to the business. The goal is to assure the business of the certainty of outcome with respect to the original business case. It covers the following three planning and delivery themes:

➤ Sponsorship.
➤ Benefits management.
➤ Business change management.

How to manage delivery of a project business plan

To track the delivery of the project business plan, a Project Manager must behave both as a consultant and a business manager in order to actively manage the business strategy (Figure 3-1).

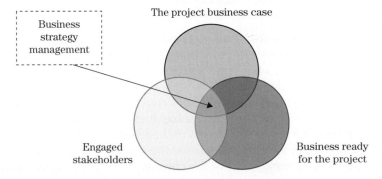

Figure 3-1 Business strategy management

The Project Manager has to consider how best to manage the agreed business strategy for that project considering all 'hard' and 'soft' aspects. Although the key relationship in delivering the business plan is with the sponsor, the Project Manager also needs to consider how the wider stakeholder group can support the delivery of the project business strategy.

The Project Manager as a business manager

The Project Manager has to spend some of his time looking outside of the project, behaving as a business manager:

➡ Working with the sponsor to manage senior stakeholders in the business.
➡ Scanning the business horizon for issues of potential impact on the project.
➡ Understanding and interpreting current business strategy as it impacts the project and as it is impacted by this project itself.
➡ Managing the delivery of the business case so that the investment reaps the required rewards.

It is too easy for a Project Manager to get lost in the project and become completely internally facing. Analysing the external environment is a fundamental part of this role and the usual business management tools can be used such as a SWOT Analysis (Table 3-1).

Short case study

A project to improve the quality control within an engineering company was in progress. The Project Manager was one of the senior chemical engineers in the company and as a part of his ongoing business management he frequently conducted a project SWOT Analysis to evaluate whether the project business case, and the associated project scope and cost/time objectives, remained valid. He would usually do this in conjunction with his small part-time team using the following process:

➡ The team would initially brainstorm the current strengths within the project – and then rank these in order of importance to project success.
➡ Then they would move on to consider weaknesses, opportunities and threats – each time identifying and then ranking.
➡ Once all areas have been ranked the Project Manager took the two highest ranking issues in each area and challenged the team to consider:
　▷ How each strength could be further strengthened
　▷ How each weakness could be eliminated
　▷ How the team could maximize each opportunity
　▷ How the team could minimize each threat

An example from one of these analyses (Table 3-1) demonstrated how a strength can be used to eliminate a weakness. In the example the strength, an active sponsor, was used to gain senior management support for the project and do so in a way which eliminated the pressure the team were feeling when completing their timesheets. The senior team were encouraged to allocate a set number of man-hours to the project and team members could use a specific project number to note their time working on project delivery. Although a budget had been previously set it was not overtly communicated so this strategy did reduce the pressure the team were feeling.

The completion of the highlighted actions supported the team reacting to its environment in a positive and proactive way and by using some of the current projects as 'pilots' the team were able to realize benefits much sooner than anticipated.

Table 3-1 Example project SWOT Analysis

Planning Toolkit – SWOT Table			
Project: Quality Control Improvement		**Sponsor:** Quality Director	
Date: Month 5 of 12		**Project Manager:** Senior Chemical Engineer	
SWOT identification and analysis			
Strength	**Ranking**	**How strengthen?**	
Active sponsorship	1	Give the sponsor an additional goal – to legitimize the project in the eyes of senior management	
Clear business driver for the project	2	Have a formal, approved business case	
A technically strong team	3		
General appreciation within the company that quality control needed to improve	4		
Weaknesses	**Ranking**	**How eliminate?**	
Pressure on project resources to prioritize 'paying work'	1	Assign a project number and a man-hour budget for each team member to use	
Using part-time resources who view this as their lowest priority	2	Communicate the validity of this project to the whole business – gain more wide-spread credibility	
The lack of robust data on current quality issues	3		
Lack of involvement of all team leaders	4		
Opportunities	**Ranking**	**How maximize?**	
Clients are going through huge business improvement exercises themselves and expect it of the suppliers they use	3		
The Project Manager has recently been asked to join an industry wide group reviewing engineering standards	2	Get approval to join this group to exchange knowledge which can be brought back to the project	
There are many new projects about to start – some are prime candidates for a different approach to quality management	1	Select one small project and use it to proactively test some of the new approaches – including new measures and ways to communicate them	
Threats	**Ranking**	**How minimize**	
Clients are complaining about quality on current projects	1	Identify the projects and use them as 'live' pilots – solve problems using the new approaches	
Other project teams see this project as a form of covert auditing and so are resisting supporting it – they do not want to be checked up on	2	Communicate the goals of this project to the wider business so that it can be seen as supporting positive change not hunting out poor performers	
There are many new projects about to start having an impact on resources and ability for the company to apply new approaches	3		
SWOT summary			
It is critical that the project remains linked to the current business priorities or it will just be seen as an 'add-on' initiative rather than a business change project. Most other improvement projects within the company have failed due to this reason. Need to ensure that the team members feel that working on this project is valid and value add			

The importance of doing a SWOT Analysis is not just in the identification of issues which might impact the project or the business but in highlighting the most important issues. For these the Project Manager should be developing proposed actions for discussion, with the sponsor, and then for implementation. There is no point in going to the effort of completing a SWOT Analysis if there are no actions resulting and nothing changes as a result.

The SWOT Table was introduced as a project management risk tool in *Real Project Planning* (Melton, 2008) and can be used within a project in a number of ways: assessing all elements of project delivery.

The Project Manager as a consultant

The term 'consultant' refers to the way a Project Manager behaves when he is operating outside of traditional project boundaries, for instance when a Project Manager is operating on the boundary between the project and the business. In this case the Project Manager is providing advice to the sponsor and other external stakeholders. In order to provide this advice the person or persons need to be in an appropriate relationship to receive it, therefore the use of the consultancy lifecycle to develop and maintain this relationship is appropriate (Figure 3-2).

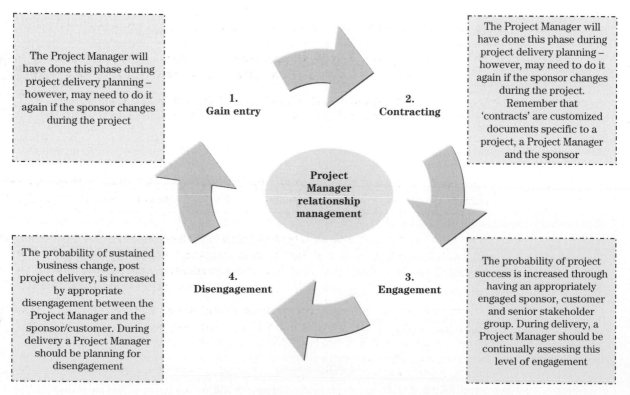

The Project Manager will have done this phase during project delivery planning – however, may need to do it again if the sponsor changes during the project

1. Gain entry

2. Contracting

The Project Manager will have done this phase during project delivery planning – however, may need to do it again if the sponsor changes during the project. Remember that 'contracts' are customized documents specific to a project, a Project Manager and the sponsor

Project Manager relationship management

The probability of sustained business change, post project delivery, is increased by appropriate disengagement between the Project Manager and the sponsor/customer. During delivery a Project Manager should be planning for disengagement

4. Disengagement

3. Engagement

The probability of project success is increased through having an appropriately engaged sponsor, customer and senior stakeholder group. During delivery, a Project Manager should be continually assessing this level of engagement

Figure 3-2 The consultancy lifecycle and business plan delivery

Step 1: Gain entry

This first stage in building an effective working relationship with the sponsor is only used during project delivery in three instances:

- The relationship has broken down and needs to be rebuilt. In this extreme situation if neither the sponsor nor the Project Manager are to be replaced then the relationship needs to be restarted. A part of this is clearing the air and understanding the root causes for the current situation.
- The sponsor is replaced and the Project Manager needs to build a relationship with the new one. This is a likely scenario due to the frequency with which senior management move roles within an organization. It can also be caused by organizational changes so that the sponsor no longer has the appropriate authority to be the sponsor.
- The Project Manager is replaced and the new Project Manager needs to build a relationship with the current sponsor. For the Project Manager this is the usual process of getting to know the sponsor, however for the sponsor this may feel like going over old ground.

Step 2: Contracting

Relationship contracts are customized for the individuals in a relationship and so if there is a new relationship being built (for one of the reasons outlined in Step 1) then a new contract needs to be agreed. It does not automatically follow that a new sponsor would have the same expectations or needs, nor a new Project Manager.

The majority of contracting is completed during project delivery planning and examples of contracts are included in *Real Project Planning* (Melton, 2008). Typically a contract with a sponsor or external stakeholder would include:

- What constitutes success in the relationship.
- What constitutes project success.
- How the two individuals will work together: meetings, communication, reporting.
- What each individual will bring to the relationship: resources, information, decisions.
- The criteria for disengagement.

Step 3: Engagement

Assuming an effective relationship has been established the most important issue during project delivery is maintaining that engagement throughout all aspects of delivery – when things are going well and when things are going badly. Engagement can be measured any number of ways (page 35), but the basis is that the two individuals in a relationship are committed to their contract and maintain it.

Step 4: Disengagement

Upon realization of the business benefits a Project Manager needs to disengage from project specific relationships. Too often relationships end prematurely without consideration of:

- Whether the contract has been delivered.
- Whether there is any need for the relationship to continue – usually linked to sustainability.

During delivery the Project Manager should already be planning for disengagement.

Sponsorship

Maintaining an engaged relationship with the project sponsor is one of the most important roles of any Project Manager. Without an active sponsor a project can become isolated and eventually completely disconnected from the business it is supposedly supporting.

Sponsorship management and engagement

Assuming an effective relationship has been built with a sponsor there is nothing more important than maintaining that relationship. However with every relationship there are forces which will support or resist its ongoing success (Figure 3-3).

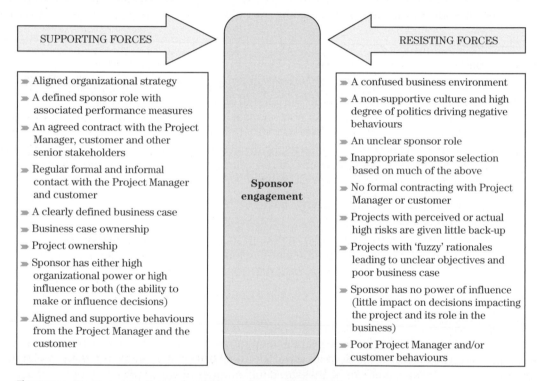

SUPPORTING FORCES → Sponsor engagement ← RESISTING FORCES

Supporting Forces
- Aligned organizational strategy
- A defined sponsor role with associated performance measures
- An agreed contract with the Project Manager, customer and other senior stakeholders
- Regular formal and informal contact with the Project Manager and customer
- A clearly defined business case
- Business case ownership
- Project ownership
- Sponsor has either high organizational power or high influence or both (the ability to make or influence decisions)
- Aligned and supportive behaviours from the Project Manager and the customer

Resisting Forces
- A confused business environment
- A non-supportive culture and high degree of politics driving negative behaviours
- An unclear sponsor role
- Inappropriate sponsor selection based on much of the above
- No formal contracting with Project Manager or customer
- Projects with perceived or actual high risks are given little back-up
- Projects with 'fuzzy' rationales leading to unclear objectives and poor business case
- Sponsor has no power of influence (little impact on decisions impacting the project and its role in the business)
- Poor Project Manager and/or customer behaviours

Figure 3-3 Forces supporting and resisting sponsor engagement

A key part of maintaining an engaged sponsor is the management of these forces. Whilst some relate to the content of the business set-up plan, others relate to the behaviour of the sponsor and their engagement with the project, Project Manager and project outcome for the business.

Defining an engaged and active sponsor

Whether a project is easy or difficult, low or high risk, simple or complex, it needs a sponsor. A sponsor is accountable for the realization of the business benefits, for delivering the required business changes and for ensuring that the project is still valid within the organizational environment.

The sponsor has a specific role during the project delivery stage particularly with respect to the business plan – the plan which links the project to the business (Table 3-2). The role of an 'active' sponsor is therefore a major undertaking and should not be confused with more traditional sponsor 'inactive' behaviours seen (Table 3-3).

Table 3-2 The role of the sponsor during project delivery

Delivery activity	Sponsor role	Project Manager role
Stakeholder management	➤ Accountable ➤ Actively involved in managing specific senior relationships such as the customer	➤ Responsible ➤ Actively involved in managing the overall delivery of the Stakeholder Management Plan and measuring engagement ➤ May delegate some stakeholder relationship management to key team members
External communications	➤ Accountable ➤ Provide general direction based on current status of external business environment ➤ Provide advice and/or approval on intended communications ➤ Provide feedback received from external stakeholders	➤ Responsible ➤ Develop key messages and propose communication vehicles based on sponsor direction ➤ May delegate some communication generation to key team members
Benefits management	➤ Accountable ➤ Maintain a high-level overview of the benefits delivery including the risks ➤ Support benefits risk mitigation	➤ Responsible ➤ Track benefits realization ➤ Monitor benefits risks and mitigate them
Business change management	➤ Accountable ➤ Deliver agreed element of the business change plan ➤ Approval of resource release for business change activities ➤ Support business change mitigation	➤ Responsible ➤ Maintain a clear link between the project scope and the scope in the business change plan ➤ Track business changes ➤ Monitor business change risks and mitigate them
Project team management	➤ Support mitigation of team capability or capacity risks through links to the business ➤ Potential mentoring of the Project Manager or a senior team member	➤ Accountable and responsible ➤ Actively manage the team and mitigate capability or capacity risks ➤ May delegate some team management to team leaders in a larger project
Project roadmap management	➤ Agree go/no go decision points during project delivery ➤ Champion approval of stage gates where other senior stakeholders are involved	➤ Accountable and responsible ➤ Track achievement of stage gates ➤ Delegate delivery of activities and deliverables to achieve stage gates to team members
Project finance management	➤ Financial governance support and advice	➤ Accountable and responsible

(Continued)

Table 3-2 (Continued)

Delivery activity	Sponsor role	Project Manager role
Project scope management	➡ Approve scope changes which have a potential impact on critical features	➡ Accountable and responsible ➡ Track delivery of value via delivery of critical features as per the project set-up plan
Project control management	➡ Approve changes to cost, scope, time which have a potential impact on the achievement of the business case	➡ Accountable and responsible ➡ Track and forecast key project control indicators as per the project control plan

Sponsor assessment

The issue of sponsor assessment is one which has been raised frequently in recent years. All too often sponsors are given this role and have little understanding of the responsibilities and accountabilities. It is therefore appropriate, as a part of the business plan engagement process (Figure 3-1) to frequently assess whether:

➡ The sponsor is meeting the needs of the project.
➡ The Project Manager is providing sufficient support to the sponsor.

Therefore three alternative methods of measuring sponsor engagement are suggested:

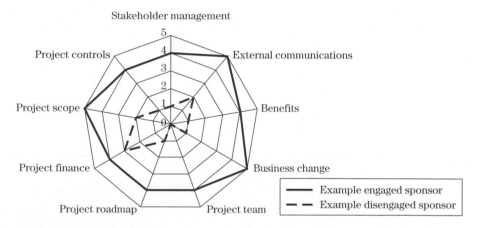

Figure 3-4 Sponsor assessment chart

➡ Method 1 – Table 3-2 can be used as a checklist against sponsor actions and a simple spider graph can be used to generate a visual indicator (Figure 3-4).
➡ Method 2 – A simpler checklist is presented in Table 3-3. This looks at sponsor behaviours as much as actions. In this format a Project Manager may note the frequency of any poor behaviours and address each either as they occur or at an agreed review point. Generally this type of checklist is looking for behaviours that confirm:
  ▷ A sponsors belief in the project.
  ▷ That the project is a priority for the sponsor.

⊃ A shared vision of success.

⊃ A shared sense of urgency.

⊃ Personal commitment from the sponsor.

➡ Method 3 – Assess sponsorship performance by reviewing the delivery of the business plan (Table 3-4) as this is the tangible output from the Project Manager–sponsor relationship. After all, if the business plan is being delivered then this is one indicator that the sponsor–Project Manager relationship is an effective one.

Overall a Project Manager can simply assess if the relationship is red, amber or green as linked to his specific needs. Whatever the assessment method or criteria, the important thing is that a Project Manager performs actions to maintain or increase engagement as necessary for project success.

In general, when there is a completely disengaged sponsor there is only one solution, assuming that the sponsor is the right person, and that is to re-contract. A Project Manager needs a sponsor to share the vision of success, believe in the benefits for the business and champion those benefits (and the resources and changes required to achieve them).

If recontracting is not possible, then the ultimate solution is to change one individual in the relationship. Assuming that the Project Manager is performing and the project is meeting current performance targets it is quite difficult to get support for a new sponsor. The impact on the project will need to be clearly articulated and communicated to senior management. Once a Project Manager has made the decision to work around the sponsor the relationship is irrevocably broken.

Table 3-3 Signs of a disengaged sponsor

Sponsor behaviour	Evidence	Potential causes
Commitment to supporting the Project Manager	Keeps cancelling scheduled 1–1 meetings	➡ This may be due to a poorly defined contract or no contracting having been done ➡ If a good contract has been developed addressing these issues then the sponsor is breaking the 'deal' and this needs to be discussed and resolved
	Does not provide decisions in a timely manner	
	Does not support removal of a disruptive and poorly performing team member	
	Does not turn up to invited project team meetings	
Commitment to supporting the organization	Cannot articulate the need for the project and associated business changes	➡ This may be because the sponsor does not believe in the project or associated business change or does not understand them ➡ This may be due to the sponsor being allocated the project from an inappropriate place in the organization or due to having too many projects to sponsor ➡ This may also be because the project is too risky and the sponsor wants to distance himself from it
	Does not frequently and publicly show support for the project	
	Does not protect the project from the political and/or organizational consequences of change	
	Has not gained support for the project and associated business changes within the organization	

Change of sponsor

For whatever reason a sponsor leaves a project, the process to engage with the new sponsor remains the same as the first time it was done (Melton, 2008). A Project Manager should:

- Ensure that a sponsor is selected according to organizational position with respect to the ultimate business change – remember that a sponsor needs to be at the lowest possible level in the organization to have the necessary authority to make decisions.
- Meet with the new sponsor to 'gain entry' and to start 'contracting' – it may be that you need to reflect on what you got from the previous sponsor and how this supports project success, or that you need to explain current project issues and how the sponsor can support resolution. During project delivery contracting with a sponsor should be very explicit and detailed.
- Manage the contract providing appropriate feedback to the sponsor on his performance as well as project progress.

Tool: Business Plan Review Checklist

Based on the ongoing importance of effective sponsorship during project delivery, a tool has been developed to support effective sponsor engagement, incorporating a review of the status of business plan delivery (Table 3-4). The Business Plan Review Checklist is intended to be used by the Project Manager in partnership with the sponsor as a method of contract review and ongoing relationship management as well as business plan delivery tracking.

Sponsor identification

It is important that it is clear who is taking organizational accountability for the realization of the business benefits and whether this has changed during the delivery of the project. Sponsors are selected based on their position in the organization and it is also important if their organizational role and authority (and potentially influence) changes. Examples of issues this may highlight are:

- The sponsor may have moved roles within the organization and no longer be the appropriate sponsor for the project.
- The sponsor may have delegated the role to a reinforcing sponsor and this needs to be assessed for appropriateness.
- The sponsor may have been replaced for a variety of reasons and the appropriateness of the new sponsor should be reviewed and a new contract developed.

Sponsor contract review

The aim of a contract review is to assess three things:

- *Is the sponsor doing his job?* – scanning the far external environment (the organization and the industry in which it works) and the near external environment (the business area which the project is supporting).
- *Is the Project Manager doing his job?* – delivering the scope required to deliver benefits at the right cost and within the required time.
- *Is the contract still valid?* – based on the above the contract may need to be changed to reflect different communication needs, changes in business drivers or issues caused by project delivery. The communications plan would normally be reviewed to ensure that it is being effectively delivered and/or that it remains appropriate.

Table 3-4 The Business Plan Review Checklist explained

Delivery Toolkit – Business Plan Review Checklist			
Project:	<insert project title>	**Project Manager:**	<insert name>
Date:	<insert date>	**Page:**	1 of 1
Sponsor identification			
Sponsor name <insert the name of the person who is accountable for the delivery of the project business case, including all business benefits> **Sponsor role** <insert the organizational role of the sponsor and also any other specified role linked to this project>			
Sponsor contract review			
Is the sponsor operating to the agreed contract? <insert the decision – yes or no – with agreed actions > **Is the Project Manager operating to the agreed contract?** <insert the decision – yes or no – with agreed actions> **Does the contract need to be amended in any way?** <insert the decision – yes or no – and detail any changes to the contract in terms of activities that the sponsor or Project Manager are to do>			
Business strategy management			
Will the business case be successfully delivered? <insert the decision – yes or no – with comments and agreed actions> **Are all stakeholders appropriately engaged?** <insert the decision – yes or no – with comments and agreed actions> **Is the business ready for the project?** <insert the decision – yes or no – with comments and agreed actions>			
Summary and action plans			
Is the progress of each area of the project business plan according to plan? <insert the decision – yes or no – with comments> **Summarize any actions needed to support successful business plan delivery** <insert any agreed actions from the sponsor/Project Manager meeting with timings, responsibilities and success criteria>			

Business strategy management

The sponsor and Project Manager should assess the probability that the business plan is being successfully delivered and focus on three main areas:

- *The business case* – assess any issues with the cost/benefit delivery, noting that the business case is a contract between the sponsor and the business.
- *Stakeholder engagement* – review the current status of the Stakeholder Management Plan and in particular any areas of engagement which are lower than plan. The majority of senior stakeholder issues will require support from the sponsor.
- *Business readiness* – review the status of the business unit into which the project will be integrated as BAU and highlight any issues, particularly those that require customer, end user or senior stakeholder involvement.

Summary and action plans

The Project Manager and sponsor need to agree on the status of the business plan at frequent stages during delivery. They must ensure that their relationship is robust and able to identify issues before they seriously impact the project and its ability to deliver the business plan and therefore business case. They need to agree any mitigating actions which may be:

➤ *Internal* – requiring the Project Manager to perform a mitigating action within the project. For example changes to scope delivery or timing of that delivery to cope with changes in the external project drivers. This would need to be documented by appropriately approved project change control requests.
➤ *External* – requiring the sponsor to perform a mitigating action external to the project within the business. For example requests for additional resources (assets, funds or people) to be released to the project. This would need to be documented by appropriately approved business resource requests.

Communication management

During the development of the PDP one of the key roles of the sponsor is to work with the Project Manager to generate a project specific communications plan to ensure appropriate communications delivery external to the project. Effective communication is at the heart of the delivery of the business plan, as it has the potential to impact the effectiveness of stakeholder and business engagement.

The communications process

In general communications delivery will involve the Project Manager taking project specific information and presenting this in different formats, using different communications vehicles, for a variety of audiences (Figure 3-5).

The key stages in an effective project communications process are as follows.

Horizon scanning

In PR (public relations) terms the term 'horizon' is used to describe the near and far environment which you are attempting to communicate with. Horizon scanning is therefore the act of reviewing those environments for potential threats, opportunities or other developments which are likely to have some impact on the project. In a project context the scanning is of the external business environment and the internal project environment. Such scanning is additional to the management of the communications plan, developed prior to commencement of project delivery and can identify:

➤ *Stakeholder engagement issues* – linked to company politics or inappropriate behaviours on topics closely linked to the project or business area impacted by the project.
➤ *Potential changes in the business* – policies or people for example – which may be opportunities or threats.
➤ *Future developments* – such as the introduction of new technology, a change in the external environment in which the industry competes or financial trends.
➤ *Project Team issues* – typically people or relationship issues which might not be noticed immediately within the project control tracking outputs (which focuses on deliverables progress).

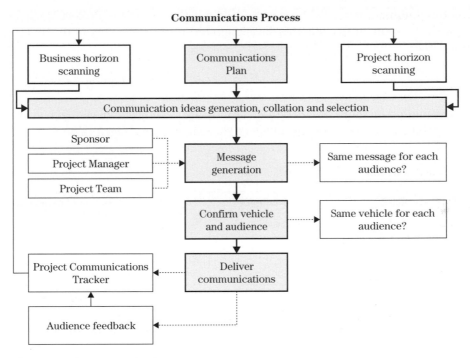

Figure 3-5 The communications process

Based on the scanning of the environment the Communications Plan is usually modified. In fact this is the main reason for horizon scanning: to ensure that the communications to be delivered match the needs of the stakeholders operating within the business and project environment. Horizon scanning can also detect trends or frequent issues and should be completed by every Project Team member as a part of their 'day job'.

Ideas collation and selection

The internal communication processes should make it easy for ideas, generated as a result of horizon scanning, to be collated and reviewed. This process should also take into account the current progress, and success, of the project Communications Plan. The selection of ideas should be based on their support of project success. Examples of ideas selected as a part of horizon scanning and communications progress tracking include:

➤ *Introducing a new audience* – perhaps a group impacted in an indirect way have been omitted from the stakeholder mapping, or there might be a new senior stakeholder involved due to a reorganization within the company.
➤ *Changing the type and/or frequency of communications to a specific audience* – it may be that a senior stakeholder is feeding back that he needs more/different information.
➤ *Developing a specific, targeted response to feedback* – perhaps an individual has highlighted a valid concern or there has been a development in the internal or external environment.

Not all ideas for communication will be selected and it is the role of the Project Manager in partnership with the sponsor to make this decision in terms of the delivery of the external Communications Plan.

Message generation

Within every communication there should be a clear message. The message should be generated in order to illicit a specific response from the audience receiving the message (Table 3-5).

Table 3-5 Messages and reactions

Example message	Audience	Anticipated response
The project is progressing according to plan	End users	We are going to move into the new facility on time so can plan accordingly
	Senior stakeholders	We can make an external announcement regarding the launch of our new product
Project delays are anticipated due to lack of resources	Senior stakeholders	We need to understand what resources are needed and release them
The anticipated risks within the business have been mitigated	Senior stakeholders	It is all under control and I do not need to do anything additional
The benefits will not be delivered	Sponsor	Understand why and support development of contingency plans if possible
The changes in the business are going to plan	End users	Motivation to continue to support the project by delivering the remaining changes
	Senior stakeholders	Continue to release resources to do these, often considered, 'non project' activities

When generating message content it is important that the likely response is forecast. Whether the 'news' is good or bad an audience will have a response – from making a decision within the business to getting involved in the project.

What is crucial is that the message is backed up with evidence – it is a real message, with accurate content based on the current project situation or forecast situation.

Vehicle/audience selection

As highlighted in *Real Project Planning* (Melton, 2008) there are a multitude of communication vehicles which can be used for a whole variety of audience types. For every message there will need to be a decision regarding who receives the message and how.

Communications delivery

The message needs to be delivered to the audience(s) as planned – both in terms of how and when. Communications delivery could involve the Project Team and end user representatives as well as the Project Manager and sponsor.

Communications management

The communications process needs to be managed effectively so that it continues to support the project. Therefore there should be a mechanism to ensure that:

- Communications occur as planned.
- Feedback is measured.
- Feedback is used to adjust the communications going forward as necessary to ensure the project continues to receive the appropriate external support.

Communications feedback

The Project Manager has to ensure that the communications process is effective both internally (within the project) and externally (within the business). He must achieve a fine balance between communicating too much and too little. The only way to do this is to get feedback on the communications. The timing and format of the feedback will depend on the communications vehicle used. For example:

- *A verbal communication* – can get an instant verbal or non-verbal response. In other words when you tell people something face to face their body language or actual response to you tells you if the message you intended has been received. In this situation the communication can be instantly amended to respond to the feedback if the message was not effectively received.
- *A non-verbal passive communication* – may need an active request for feedback at some time after the communication has been made available. For example a survey might be issued to staff to see if they have been using a new notice board and what benefits they believe this provides. Any survey based feedback needs to bear in mind that this will not get a 100% audience response, however the response rate needs to be significant enough to be used as an indicator of the feedback from the whole audience population.

The Project Manager and sponsor need to agree how feedback should be obtained so that they have reassurance that communication is appropriate and supporting the project and the business.

Internal project communications

The basic principles of communication management and the communication process are the same no matter if it is internal or external to the project (internal Project Team communications are covered in Chapter 4). However, in terms of managing the business plan, the Project Manager and the sponsor must ensure that key messages from the external environment are communicated to the Project Team as necessary. For example, if a change in corporate purchasing policy has occurred, then the Project Team delivering local purchasing process improvements needs to be made aware of this.

Short case study

The following short case study demonstrates an extreme case of what can happen when there is no focused project communications process, leading to inappropriate communications between the project and the business and the organization's external environment.

During a new product launch project the senior marketing executives had separately communicated with the shareholders and the external marketplace on the anticipated launch date. The Project

Manager had not been involved in this decision and in fact the sponsor and his business management team did not consider such matters anything to do with the project.

However the project was always highlighted as a moderate to high risk and once the advertised launch date was factored into the current progress review it quickly became clear that this date was not achievable.

Both the Project Manager and sponsor had failed to identify key messages which the other needed in order to deliver the business plan.

- The sponsor should have understood the risks in deciding on a launch date without confirming all aspects of the product would be available for launch – product tested, supply chain developed and stocks manufactured.
- The Project Manager should have understood that the sponsor needed a frequent assessment of the likely date for launch based on the delivery progress for the critical aspects of scope.
- The Project Team needed a process to communicate with the Project Manager on the critical aspects of scope so that issues such as product development and testing issues (which caused problems) were highlighted earlier.

The product was eventually launched 2 months later than expected by the marketplace and by this time an adverse impact on the company share price had been seen. This reflected the concerns felt by shareholders and other investors – the company had not delivered a key new product when it said it would. As a result the overall business plan was not delivered due to reduced sales and reduced financial benefit.

Tool: Communications Tracker

Based on the need for an effectively delivered communications process, a tool has been developed to support communications management (Table 3-6).

Table 3-6 The Communications Tracker explained

Delivery Toolkit – Communications Tracker				
Project: *<insert project title>*		**Project Manager:** *<insert name>*		
Date: *<insert date>*		**Page:** *1 of 1*		
Communications goal				
<insert SMART measures taken from the original Communications Plan – usually related to achievement of stakeholder management objectives and also linked to support of the project vision of success>				
Key message	**Audience**	**Activity**	**Feedback**	**Mitigating actions**
<insert the key message of the communications content>	*<insert the category of stakeholder or each individual stakeholder>*	*<insert the specific method by which the key message will be communicated>*	*<insert how successfully the key message was received and the mode of feedback>*	*<insert any mitigating actions as a result of the feedback>*
Communications summary				
<insert summarizing comments on the effectiveness of the Communications Plan>				

The Communications Tracking Tool is intended to be used by the Project Manager in partnership with the sponsor as a method of reviewing the effectiveness of communications and in particular the success of the external Communications Plan.

Communications goal

The overall aim of the communications from the project to external audiences should be articulated and clear measures used to demonstrate that it is effective and meeting the needs of the entire stakeholder group. Example communications goals are:

Zero complaints regarding communications

Although this is a negative measure it is important as it can focus the team on ensuring that the right level of communications is maintained for each stakeholder type. Too much pointless communication can illicit as many complaints as too little communication.

Greater than 80% engagement of important/influential stakeholders

Communication is one method to engage with senior stakeholders. Their behaviours are often linked to the way that information is fed to them. If the right level of information is available then they are able to make appropriate decisions.

Greater than 80% of end users feel informed about the project

A key influential stakeholder group is the end user, those who will interact with the project scope after handover and integrate it into BAU. When a group is going through any form of business change the frequency, type and tone of communication can assist in removing resistance, or, the converse.

Zero adverse stakeholder behaviours, actions or decisions

When stakeholders act in a way that is detrimental to the project it can be because they are doing so in ignorance. In other words they do not have the information to behave, act or make decisions which support the project and/or business.

Key message

The specific key message being communicated should be articulated. For example:

- *Progress update* – positive news, project on plan.
- *Issues update* – negative news and mitigating actions.
- *Outcome of the design phase* – key changes and impact on end users.
- *Likely timing of implementation* – confirming when end users will be impacted.
- *Benefits update* – positive news, linked to business needs.

Audience

Typically the audience will have been categorized during communications planning into various stakeholder groups. Not all key messages are communicated to all audiences and sometimes the message may be slightly different for different audiences because of the difference in reaction/action needed from them.

Activity

There are many types of communication activity and it should already have been agreed within the Communications Plan which is the most appropriate to ensure that the right message is received by the audience. However as the project is delivered there will be a need to deliver additional unplanned communications and/or to reconsider a previously agreed vehicle for a specific message. The majority of communications vehicles would fit into the following four categories.

Written–visual

Physical, visual communications are appropriate where there is need for a continuing reinforcing message through the audience having continuing access to the message. For example the use of a notice board at a restaurant entrance, the distribution of newsletters and access to them within all communal areas of an office complex, the use of posters in specific areas with messages targeted for staff working in those areas such as safety messages on a construction site. Within any one project care needs to be taken in using this format excessively for communications – too much and the reinforcing value may be diminished as people stop 'seeing' the messages.

Written–electronic

Electronic communications via personal email or intranet systems are an effective way to get into someone's desk area – where they are working. It is also a way of encouraging active communication with the audience through having an electronic area where they can go to for updates/information. However, caution should be used with too much email (particularly non-focused) to a large audience. With the increase of email, particularly junk email, people are getting very selective with what they will actually read.

Verbal–individual

Personal one to one sessions are appropriate for very customized messages – developed and delivered for an audience of one. It allows the person delivering the message to get immediate feedback – both verbal and through audience body language. Often 1–1 sessions are used to communicate the impact of a project on a particular end user group and a standard message can be used with variations for each team member. However caution should be used with this method of communication for all but the clearest messages. If the communicator is not careful the message can be incrementally changed as each team member 1–1 session generates questions, so that the final 1–1 might convey quite a different message from the first.

Verbal–group

Group sessions are an excellent method to communicate a single consistent message to a large group. Some feedback via question/answer sessions and body language can be gained, however the 'group' setting will restrict some of the audience and encourage others. In particular the communicator needs to ensure that the message will not aggravate a few resistors who then can change the nature of the session impacting those who are either neutral or positive to the message.

Feedback

There will be some expected outcome from the communications activity. For example:

- A change in the behaviour of a specific audience.
- Verbal feedback – either solicited or unsolicited.
- A completed action by the specific audience.

At various stages in the communications process there will also be a need for audience feedback to be requested and again this can be done in various ways. Surveys and focus groups are common ways to get a large volume of feedback.

Mitigating actions

As a result of the immediate or follow-up feedback, specific mitigating actions may need to be taken to:

- Maximize the opportunity if there has been a very positive outcome.
- Minimize the threat is there has been a neutral or negative outcome.

Communications summary

The Project Manager should be able to assess the effectiveness of the delivery of the Communications Plan in terms of the cumulative effectiveness of messages received by the intended audiences. This should be measured by tracking progress of the initial measures defined in the communications goal which usually relate to the overall engagement of a particular audience.

Short case study

A project to review and relocate a technical services department is in progress. The project had only recently developed a structured Communications Plan (during Month 5), yet the positive benefits of delivering clear, appropriate messages were starting to be seen. The project involves a series of major changes which will impact all the staff within the department and effective communication was identified as a critical element. Table 3-7 is an extract from the monthly Communications Tracker (Month 7).

This case study demonstrates how poor communication can exacerbate an already fraught situation: the design of the new facility had been done by the Project Team with little external participation from the end users, although their views had actually been integrated into the new ways of working they weren't aware of this. In addition the staff relocation had to be managed within company policies and procedures which restricted too much early communication.

It equally demonstrates that appropriate communication at the right time can turn a situation around: In this example, the staff became less resistant once they could understand the personal impact of the changes.

Table 3-7 Example Communications Tracker

Delivery Toolkit – Communications Tracker				
Project: Technical Services Review		**Project Manager:** Kath Plummer		
Date: Month 7 year 1		**Page:** 1 of 1		
Communications goal				
The technical services teams are motivated and engaged in the changes included within this project. Specific measures are: retention of staff during the project, level of absenteeism, sickness, disciplinary and grievances. Apart from HR-related measures the level of participation of team members in the project will be tracked.				
Key message	**Audience**	**Activity**	**Feedback**	**Mitigating actions**
New facility design eliminates issues with the current facility	Technical Services Team	Group meeting with facility layouts/ pictures available	Lots of questions. People seem to understand that the design is complete and comments appear to be mainly how they will work within the new facility	Keep photos/layout on a notice board in current working area with a space for additional comments
New facility location	Technical Services Team	Two scheduled visits to the site of the new facility followed by 1–1 sessions	Visits seemed to alleviate some concerns and 1–1's went better than expected. Some still believe travel to the new location will be a problem but not as many as initially forecast based on earlier 1–1's	Follow-up 1–1's with those who have recently changed to a positive stance as well as those still resisting the move
Communications summary				
Absenteeism was trending upwards as the project commenced delivery and reached an all time high during month 6. However the communications delivered in month 7 appear to be having a positive impact. Of the initial 30% of staff who were resisting the relocation, half are already engaging with the move. Absenteeism is forecast to reduce in coming months in line with this change in morale and engagement with the move.				

Benefits Management

The management of benefits is a crucial project process which is typically managed by the Project Manager on behalf of the sponsor. In *Project Benefits Management* (Melton et al., 2008) the overall process of the management of benefits from concept to realization was described with associated tools (Figure 3-6).

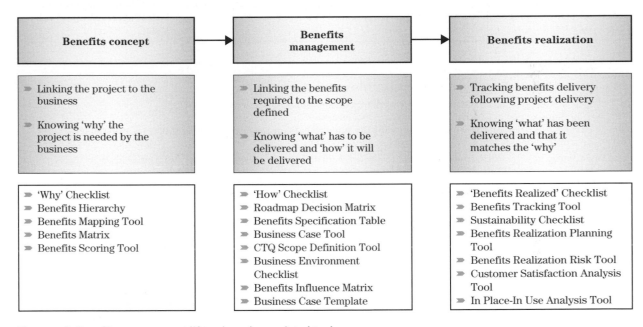

Figure 3-6 Benefits management lifecycle and associated tools

Some of the tools highlighted in Figure 3-6 are basic tools, such as the 'Why' and 'Benefits Realized' Checklists (Melton, 2007 and Appendices 9.1 and 9.4), whilst others are more advanced in terms of the skills and techniques to use them.

What the lifecycle is attempting to convey is the continual interaction and integration of project and benefits management.

The benefits baseline

Depending on the type of project the benefits baseline may have been integral to the business case or PDP development or, in rare cases, may be found during initial project delivery. The benefits baseline is the point from which the realization of business benefits will be assessed (Table 3-8).

Most projects do not have one simple benefit baseline. The business case is usually linked to a balanced view of organizational impact such as that proposed by the Balanced Scorecard (Kaplan and Norton, 1996) (Figure 3-7). Therefore it is usual to consider the cumulative impact of a project in terms of total benefits across these four areas of an organization as shown in the Benefits Totalizer tool (page 49).

Table 3-8 Benefits baseline examples

Project type	Benefit criteria	Benefit metric	Benefit baseline
Business improvement	Financial savings	Operational efficiency increase	Total number of production man-hours per kg of product
Product launch	Financial profit Meeting customer needs	Market penetration	Market share
Capital engineering	Compliance Security of supply	Manufacturing capability	Total number of defects and stoppages in any production shift

Based on the Kaplan and Norton Balanced Scorecard approach

➤ A strategic management system – driven from organizational vision and strategy
➤ Emphasizing both financial and non-financial measures
➤ Assists the organization in understanding the strategy and how they can positively support it, generating appropriate action (such as alignment of annual objectives)

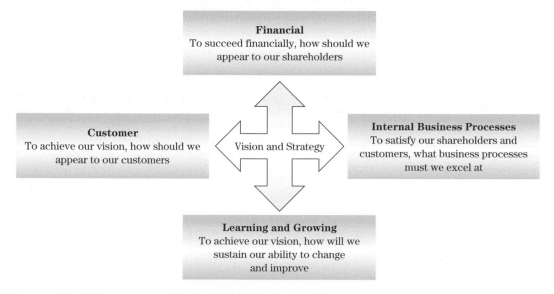

Figure 3-7 The balanced scorecard methodology

Short case study

A business change team are delivering a project to improve performance in terms of their effectiveness in delivering appropriate business changes for their organization. Due to a lack of existing performance measures the business case focused mainly on business change project speed. As delivery progressed performance data was collated and the true benefits baseline identified, allowing a clearer definition of benefits targets (Table 3-9). This gave the improvement project a more balanced target.

In order for the improvement project to be successful all benefits needed to be progressed so that they either achieved their target or significantly progressed to wards it.

Table 3-9 Example benefits baseline and totals

Benefit type	Benefit criteria	Benefit metric	Baseline benefit	Revised target
Financial	Operational efficiency	Change team utilization	60%	80%
Business process	Operational effectiveness	Projects progressing as per plan	50%	90%
Customer	Business change speed	Project delivery cycletime	35 weeks	22 weeks
	Business change robustness	Sustained benefits delivery	50%	90%
Organizational growth	Change team capability	A combination of the above	Zero projects delivered	Two successful projects/year

Tool: Benefits Totalizer

During the delivery of the project, when it is too early for detailed benefits metrics tracking, the Project Manager in partnership with the sponsor should continue to take an overview of the benefits being delivered by the project. The aim of the Benefits Totalizer (Table 3-10) is to confirm a high level forecast of benefits delivery.

Table 3-10 The Benefits Totalizer explained

Delivery Toolkit – Benefits Totalizer					
Project: *<insert project title>*			**Project Manager:** *<insert name>*		
Date: *<insert date>*			**Sponsor:** *<insert name>*		
Customer		**Rating**	**Business process**		**Rating**
<insert high level customer total benefit as a result of this project>		*<insert RAG rating>*	*<insert high level business process total benefit as a result of this project>*		*<insert RAG rating>*
Organizational		**Rating**	**Financial**		**Rating**
<insert high level organizational benefits as a result of this project>		*<insert RAG rating>*	*<insert high level financial benefits as a result of this project>*		*<insert RAG rating>*
Summary					
<insert comments on the overall impact of the RAG rating in each area on the achievement of the business case>					

Total benefit

Each individual benefit metric or criteria should be categorized within each of the four benefit areas: customer, business process, organizational and financial. Once all metrics have been assigned a category, an overall description of the resulting benefit in each area can be made.

RAG rating

The RAG rating is a way to assess the forecast outcome for each total benefit. It is based on a traffic light system:

➤ R = red – the current forecast is that the total benefit is not going to be delivered.
➤ A = amber – the current forecast is that there are major risks to the total benefit being delivered.
➤ G = green – the total benefit is forecast to be delivered as planned.

Customer

These are the benefit criteria or metrics which demonstrate that the organization's customers are receiving some benefit from the project, for example:

➤ Increased customer satisfaction.
➤ Decreased lead-times.
➤ Increased quality.
➤ Decreased price.

Business process

These are the benefit criteria or metrics which demonstrate that the organization's internal business processes are receiving some benefit from the project, for example:

➤ Increased operational effectiveness.
➤ Decreased cycle times.
➤ Increased process robustness.

Organizational

These are the benefit criteria or metrics which demonstrate that the organization will grow and learn as a benefit of doing the project, for example:

➤ Increased people capability.
➤ Corporate knowledge generation.
➤ Improved morale.

Financial

These are the benefit criteria or metrics which demonstrate that the organization will benefit financially as a result of doing the project, for example:

➤ Increased operational efficiency.
➤ Decreased inventory.
➤ Reduction in staff.

This tool is similar in concept to the Benefits Scorecard introduced in *Project Management Toolkit* (Melton, 2007) but rather than merely reporting the status the Benefits Totalizer requires a Project Manager to forecast the likely realization potential in each area. In some respects this will report problems as a precursor to a more formal and detailed benefits risk assessment (page 25), where mitigation needs to be considered (Melton et al., 2008).

Summary

Once the totalizer has been completed the Project Manager should discuss with the sponsor and interpret the results in terms of the impact on the business.

Benefits risk assessment

During project delivery, as the project is achieving specific objectives and completing delivery of specific areas of scope, the Project Manager should be conducting benefits risk assessments. A variety of techniques are available:

- *FMEA* considers benefits failure modes such as in the Benefits Realization Risk Tool in *Project Benefits Management* (Melton et al., 2008).
- *Simple risk table* – considers risks to benefits realization, their impact and likelihood of occurrence such as in the Risk Table and Matrix in *Project Management Toolkit* (Melton, 2007).
- *Simple risk checklists* – considers specific precursors for benefits realization such as in the 'Benefits Realized?' Checklist (Appendix 9.4) in *Project Management Toolkit* (Melton, 2007).
- *Fault tree analysis* – considers the logic inherent in the benefits realization process. This technique is used within the Benefits Delivery Fault Tree (Table 3-11).

Whatever the chosen technique the Project Manager needs to work with the sponsor in order to forecast the benefits outcome. This will strongly depend on the link between the project scope and the business benefits, after all the project scope enables the benefits and its non-delivery is one cause for lack of benefits realization.

Tool: Benefits Delivery Fault Tree

The aim of this tool (Table 3-11) is to determine the probability of the non-delivery of benefits. It is based on the fault tree analysis methodology (Figure 3-8). Fault tree analysis diagrams are commonly used to demonstrate events that might lead to a failure so the failure can be prevented. These diagrams are generated through performance of the following four steps:

- Step 1 – Select the top event (an undesired event or fault). This becomes the top of a logic tree.
- Step 2 – Describe each situation that could cause that event and link it to the top event. This is usually a logic statement with the probability of failure noted.
- Step 3 – Use event and gate flow charting symbols to support the analysis so that the fault tree builds the logical process that might lead to failure. The bottom of the logic tree are the initiating events.
- Step 4 – Identify the shortest/most likely credible route from fault to initiating event and mitigate to prevent the failure.

Fault tree analysis does have a flow charting symbol convention to support analysis however this tool only incorporates the two basic gates (AND and OR) as shown in Figure 3-8.

In the example in Figure 3-8 the event which could lead to a failure in benefits delivery (that had not previously been considered) was the current level of capability in a change projects team which were themselves undergoing change. The identification of this event changed specific approaches for implementation and added additional scope (capability assessment and development) which would not have otherwise been considered.

Table 3-11 The Benefits Delivery Fault Tree explained

Delivery Toolkit – Benefits Delivery Fault Tree			
Project:	*<insert project title>*	**Project Manager:**	*<insert name>*
Date:	*<insert date>*	**Sponsor:**	*<insert name>*
Fault tree analysis			
<insert fault tree diagram – example in Figure 3-8>			
Fault tree summary and mitigation plan			
Shortest credible route		**Initiating event mitigation**	
<insert description of route from top benefits event failure to the initiating event with assessed probabilities>		*<insert any agreed actions from the sponsor/Project Manager meeting with timings, responsibilities and success criteria>*	
Current level of benefits delivery risk		**Additional mitigating actions**	
<insert low, medium or high – with comments>		*<insert any agreed actions from the sponsor/Project Manager meeting with timings, responsibilities and success criteria>*	

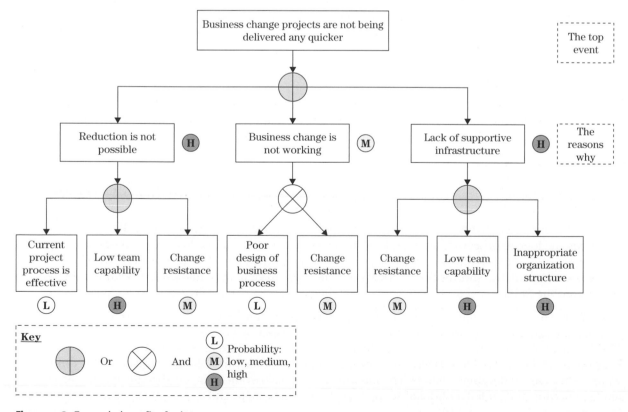

Figure 3-8 Example benefits fault tree

Fault tree analysis

Select the top event from a review of the anticipated benefits realization profile. This should be the benefits scenario which appears to be most at risk. In order to fully construct a credible and useful fault tree, as outlined in the 4-step process on page 25, the logical process to develop a positive outcome in the top event needs to be understood.

- Events which cause the top event are identified. Some have a one to one link and others require a combination of events to achieve it. Identify the probability of occurrence (low, medium or high).
- Lower level events may appear in a number of the logical branches.
- Each branch is referred to as a fault branch which identifies scenarios causing benefits failure.
- The lowest level event is called the initiating event.
- The probability of each branch causing failure of the top event can be calculated.
- The total number of events to cause failure of the top event can be counted for each fault branch.

Fault tree summary and mitigation plan

Shortest credible route

This is determined by one of two routes:

- The lowest number of events from initiating event to the top event.
- The route with the highest probability.

Generally the routes that have the highest probability of occurrence are the ones that have the fewest events. For example in Figure 3-8 there are two routes of equal credibility although both have the same initiating event (Low team capability).

Initiating event mitigation

Develop an action plan to eliminate the initiating event. If this event is less likely or cannot occur then the logical sequence of events leading to failure cannot occur. In the case of the example in Figure 3-8, the mitigating action for the initiating event (low team capability) was a training and mentoring programme.

Current benefits delivery risk

An overall assessment of the level of risk to the achievement of the top benefits event should be made. This is based on the overall fault tree diagram and the number of branches.

Additional mitigating actions

Summarize any additional mitigating actions needed to support successful benefits delivery. These may be linked to branches other than the shortest credible route, potentially the next shortest. In the case of the example in Figure 3-8, additional mitigating action for other initiating events (inappropriate organization structure and change resistance) were also developed.

Benefits tracking

Benefits realization can start the day a project starts or only the day it finishes. This depends on the project type. For example:

- *Asset projects* – are based around the integration of a new asset into a business and the project benefits are usually only enabled at the end when the asset is fully available to the business.

● *Business improvement projects* – are based around changing the business. The very act of starting to look at business performance in a specific area can start to impact both the people and the business processes causing benefits to be realized right from the start of the project. Some of these early benefits may be lost if the project does not develop a sustainable solution.

The tracking and reporting of benefits realization should align with the Benefits Realization Plan using appropriate tools and formats (Melton, 2008; Melton et al., 2008). As with the tracking of all metrics it is important that the benefits realization profile is reviewed versus plan, any gaps assessed, root causes identified and mitigating actions put into place. Figure 3-9 shows two examples for tracking benefits, a spider diagram and a data tracking chart.

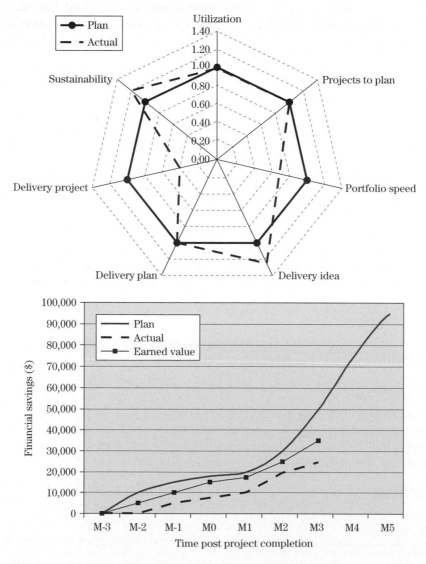

Figure 3-9 Examples of benefits tracking charts

Business Change Management

The management of business change is another crucial project process which is typically managed by the Project Manager on behalf of the sponsor. Although in some organizations the business changes are delivered and managed by the end users, the business. Either way the link between project scope delivery and business change delivery needs to be robust (Figure 3-10).

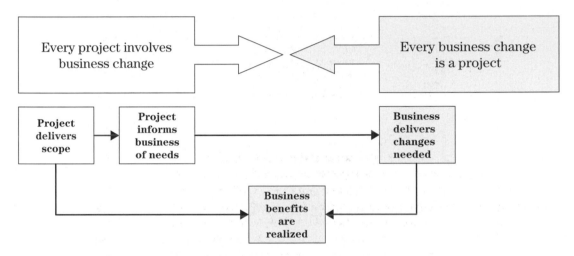

Figure 3-10 Business change and project delivery

In terms of a successful project the business change must be:

> *Planned* – so that it is the 'right' business change implemented in the 'right' way.
> *Managed* – so that it is delivered as needed by the project scope, mitigating all risks to its failure.
> *Sustained* – so that it continues to support the realization of benefits long after the project has been completed.
> *Capable of delivering benefits* – both related to the project benefits case and the wider organization.

However, successful business change also needs to be the 'right' change for an organization as seen by the stakeholders impacted or influenced by the business change. Therefore stakeholder engagement is a key part of the change delivery process.

Business change management process

Change planning

Successful business change requires significant planning to ensure effective integration within the project and then later within BAU, and this has been covered in *Real Project Planning* (Melton, 2008). Therefore at the start of business change delivery a Project Manager should have a robust business

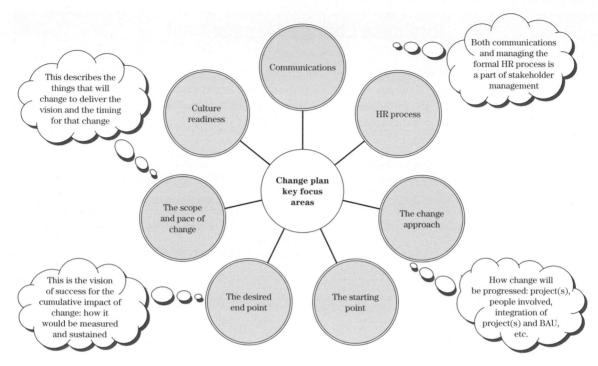

Figure 3-11 A business change plan

change plan in place (Figure 3-11). This document is a subset of the PDP which will have already identified any changes outside of the project scope which the business needs to deliver. For example:

- A revamped facility may require staff to be trained, a new facility may require additional staff to be recruited and a newly automated plan may require fewer staff.
- A new equipment item may require additional maintenance.
- A new product launch may require a change with in the overall supply chain.

Whether the scope of the business change plan is inside or outside of the scope of the project it needs to be tracked. It has to be successfully delivered in order for the overall project business benefits to be sustainably realized.

Change management

At the heart of any business change there are people: people may lose their jobs, have their jobs significantly changed or be working in an entirely different environment after the project has been delivered. Therefore the focus of business change implementation is the elimination of any people issues as the change scope is delivered (Figure 3-12).

This is managed through three key processes:

- **Sustainable change design and delivery** – designing the new way of working and putting any changes in place in a way that the changes are sustained.
- **Stakeholder management** – ensuring that the stakeholders are ready for the change and are engaged in the project outcome.
- **Risk mitigation** – ensuring that the change needed by the project is completed by mitigating all risks.

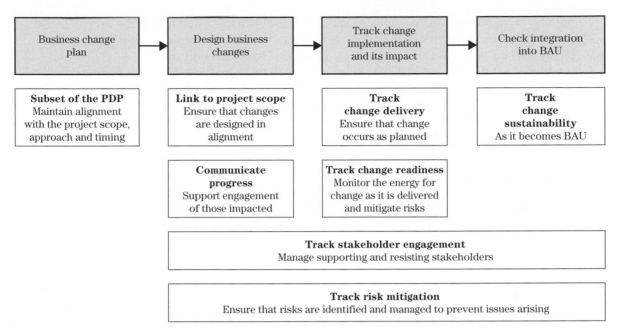

Figure 3-12 Delivering business change

The goal of business change management is to design and implement change which is supported by the business and which, when integrated with the project scope, enables sustainable business benefits.

Sustainable change design and delivery

Projects are ways to make step changes within a business and therefore a successful project is one which is eventually integrated into the business so the changes become BAU. For example, a project to buy and test a new asset can only release value to the business if:

- The business knows how it can use the asset (increase capacity, new product, new process).
- The business is ready for the asset (facility ready, new business processes, training).
- The environment is ready (the people working within the business accept the changes and integrate them into their ways of working).

A lack of business change sustainability will impact the sustainability of benefits realization. Effective business change will usually include participation by the end users so as to build energy for the change. However the important factor is that the change is designed appropriately using the 'right' design tools. For example a project which needs:

- A new organizational structure would require organizational design tools.
- Additional BAU training would require a specific training programme and trainers (either internal or external to the company).
- To improve ways of working would require specific improvement tools and techniques such as lean six sigma as well as people trained in their use (either internal or external to the company).
- To develop or launch a new product would require specific supply chain modifications or changes to external suppliers.

It would be usual for the business change scope to have been identified during the project scope development as a part of understanding how value was to be managed and delivered (Chapter 4, page 110), but it may be delivered either within or external to the main project.

Tools which track project scope delivery can be used to track business change scope delivery. These should already be detailed in the sustainability plan, another subset of the PDP. A key tool used to assess the potential for sustainability failure is the Sustainability FMEA (page 60). During project delivery this tool tracks potential failure modes which may occur during change design, implementation or at project handover and completion.

Stakeholder management

During project delivery the vast array of stakeholders, and their expectations, need to be managed. This involves working out what stakeholders care about and then responding to that. Examples of these are as follows:

- Business results.
- Happy customers.
- Happy employees.
- Quality – in the company processes or products.
- Schedules.
- Budgets.
- How their resources are used.
- What is likely to happen – the big picture.
- How they will be personally impacted.
- If/how they are involved in the project.
- How they have to change.

What an individual or group care about does depend on their position in or external to the company and/or the project:

- *Internal stakeholders* – the usual term for those within the Project Team, whether they are internal to the company or from an external supplier.
- *External stakeholders* – the usual term for those not within the Project Team but who have either authority, influence or are impacted by the project or its outcome.

Therefore stakeholders need to be managed so that they:

- Use their power and influence in support of the project and any associated business change.
- Deliver any critical resources, information or decisions.
- Believe that their expectations have been met.

During delivery the level of stakeholder engagement should be tracked. This may mean measuring any number of variables depending on the specific stakeholder and situation. For example:

- Getting direct feedback on key documents such as the Project Charter.
- Asking team members to note behaviours on a stakeholder observation chart for critical stakeholders. This would usually be a confidential document only used by a small number of team members.
- Formal stakeholder surveys, focus groups or 1–1 meetings.
- Observation 'on the job' as they deliver their commitments for the project and associated business change.

The Stakeholder Tracker (page 63) is one way to ensure that the Stakeholder Management Plan is being delivered and that engagement is at an appropriate level for success. A key point in delivery is when a significant change is about to take place, such as handover or start-up of a new asset or 'go live' for a new business process. At this stage it is usual to conduct a stakeholder change readiness review. Such a review involves checking key areas of business behaviour (Table 3-12) so that any gaps or risks to the success of the change are eliminated. The checks can be scored as yes or no, or in more complex situations a range of scores might be appropriate, for example:

- Business objectives may be fully aligned, partially aligned or not aligned.
- Supportive behaviours may be seen in all parts of the organization, only in specific parts or not seen at all.

In certain scenarios it may be better to delay business change rather than implement it when the organization is not ready. This does delay project completion and therefore benefits realization but in the long term may be the right response for the business.

Table 3-12 Signs of change readiness

Business behaviour	Change ready?	Action	
		If yes	**If no**
1. Current business objectives align with the project outcome	Yes or no	Communicate to the Project Team and the business to support wider change readiness	Work with senior management to get alignment between the project and current business priorities
2. All enabling activities are in place	Yes or no	Give recognition to those who have completed their activities	Sponsor and Project Manager need to use influence to get stakeholders to complete the necessary activities
3. All people within the organization are aware of the project and its impact on them	Yes or no	Keep communications informal but current so that everyone is completely up to date	Sponsor needs to use influence to get communications channels opened. This may also need to link with formal HR processes if the project impacts roles
4. All business communications make appropriate reference to the project	Yes or no	Continue to provide timely and accurate information to the business so that this can continue	Project Manager to provide additional information and use influence to get this integrated into appropriate channels. May need sponsor support
5. Signs of overt resistance have been eliminated	Yes or no	Keep communications informal but current so that everyone is completely up to date	Sponsor and Project Manager need to work with senior stakeholders to address specific areas of resistance
6. Senior stakeholders are behaving in an overtly aligned way – both verbally and in body language	Yes or no	Reward aligning behaviours to reinforce them. Maintain good communications at all levels	Sponsor will need to work with any resisting senior stakeholders to understand and eliminate the resistance where possible
7. All parts of the organization are behaving in a supportive manner	Yes or no	Reward aligning behaviours to reinforce them. Maintain good communications at all levels	Project Manager will need to work with resisting areas of the business to understand and eliminate the resistance where possible

Short case study

A project to launch a new product into a critical market was ready to be handed over from development into production. The production team had to significantly change the orientation of equipment and the teams that operated them as well as consider a different shift pattern. At the pre-handover meeting it was clear that:

- The equipment reconfiguration had not been completed due to the late use of much of the existing equipment in order to manufacture the final batch of another product.
- The HR issues regarding changes in roles and shift working had not significantly advanced due to additional time needed in consultation with unions.

Therefore the sponsor got agreement from the vice presidents of development and production that the launch would be delayed until an effective and sustainable handover could be achieved. This was a joint decision to delay entry into the market but was sensible considering the risks of premature entry. When the product was launched 2 months later than planned the sales exceeded expectations but the production team were more than ready to cope with this due to the successful changes within their area.

Risk mitigation

The majority of business change management activities incorporate effective risk mitigation methodologies. Too often a successful project turns into an ineffective business change due to a lack of linkage between the business change and the project outcome. Many risk tools can be adapted for use in business change risk assessment, such as the FMEA methodology, the Risk Table and Matrix and simple risk decision flowcharts. Often very simple techniques can tell a Project Manager the direction business change risk mitigation needs to take. For example, a 'what if?' analysis is a good structured brainstorming tool. This methodology can be used to build realistic scenarios against a specific business change activity (Figure 3-13). By looking at the worst case scenario and the scenarios leading to this risk, mitigation activities can be appropriately focused and prioritized.

Another useful methodology is the risk decision flowchart. This can be used to assess whether the use of resources for risk, mitigation is sensible or whether something completely different needs to be done. The Business Change Risk Mitigation Matrix (page 66) uses this methodology.

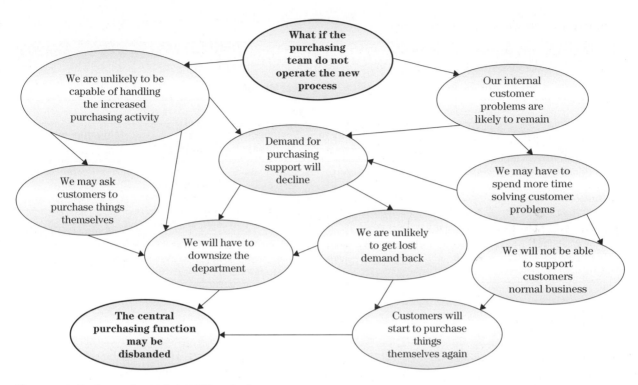

Figure 3-13 Business change 'what if?' analysis

Tool: Sustainability FMEA

The aim of this tool (Table 3-13) is to provide the Project Team with a tool to assess areas which are often outside of their direct control. It is a risk assessment tool which aims to consider all the potential sustainability 'failure modes' – those scenarios which will prevent the full and sustainable realization of the project benefits. The FMEA methodology is appropriate to use in this instance due to its inclusion of the assessment of failure mode detection. The assessment of the risk of business change and therefore sustainability failure is an important element of the project delivery process. If the risks are not properly understood, the ability to deliver benefits will be significantly affected as the appropriate sustainability mitigating actions will not be progressed at an early enough stage to make an impact on the outcome.

Table 3-13 The Sustainability FMEA explained

Delivery Toolkit – Sustainability FMEA						
Project:	<insert project title>			**Project Manager:**	<insert name>	
Date:	<insert date>			**Page:**	1 of 1	
Business change area	**Probability of not**	**Impact of not**	**Ability to detect**	**Sustainability threat**	**Risk priority**	**Mitigation plan**
<insert business change area or activity>	<insert score>	<insert score>	<insert score>	<insert score>	<insert calculated score>	<insert mitigation plan>
Scoring system						
Probability	1 = low 5 = high	Impact	1 = low 5 = high	Detection 1 = high 5 = low	Threat	1 = low 5 = high

Business change area

Each specific business change associated with the project should be listed. These may be expressed as a business change concept or a business change activity depending on the timing of the risk assessment. All business changes, whether explicit or implicit, should be included. An explicit business change is one which is defined within the approved business case, whereas an implicit business change is one which is implied through the delivery of the project. For example, an explicit business change activity associated with a project introducing a new operating process may be operator training, whereas the implicit business change might be improved team work within the operator team. Both changes to BAU are required in order to maximize benefits delivery.

Probability of not achieving

For each business change area or activity there should be an assessment of the likelihood that the business change will *not* be achieved. In order to assess this aspect a failure mode will be considered for each business change and the potential of that failure mode occuring will be reviewed. A scoring system should be defined and would typically be a 1–5 rating:

➡ 5 = high – there is a strong probability that the failure modes preventing this business change from being delivered will occur.
➡ 1 = low – there is a small probability that the failure mode preventing this business change from being delivered will occur.

Impact of not achieving

During the development of the business plan (within the overall PDP) there will have been an assessment on the relative importance or weighting of each identified business change, and therefore the impact of *not* achieving it. A scoring system should be defined and would typically be a 1–5 rating:

➡ 5 = high – there is a serious impact on the achievement of the approved PDP.
➡ 1 = low – there is a minimal impact of the achievement of the approved PDP.

Ability to detect early

It is important to understand how early in the project the failure mode could be detected. There should be a clear causal link between the business changes required and the benefits to be delivered. The Project Team should have developed sustainability checks within the PDP which track and highlight business change failure. However some sustainability checks will not be in place until completion and handover. A scoring system should be defined and would typically be a 1–5 rating:

- 5 = post-completion – the failure of the business change activity delivery (and associated sustainability check) and any associated failure mode cannot be detected until after project completion and handover. In simpler terms this can be scored as a low ability to detect.
- 2, 3 or 4 = during project delivery – the failure mode can be detected at some stage in project delivery. In simpler terms this can be scored as a medium ability to detect.
- 1 = early – the failure mode is built into the planning stage and business change activity is easy to forecast based on the planning and associated sustainability checks. In simpler terms this can be scored as a high ability to detect.

Sustainability threat

It is important to understand how business change can be sustained and the likelihood that it cannot be sustained. A scoring system should be defined and would typically be a 1–5 rating:

- 5 = high – there is a high threat to sustainability.
- 1 = low – there is a low threat to sustainability.

Risk priority number

A risk priority number can be calculated by multiplying the probability, impact, detection and sustainability ratings. A high score indicates a high risk of not delivering the business change and should therefore be considered a high priority in terms of developing an appropriate mitigation plan. The individual scores will show where the effort for mitigation needs to be focused. The highest scores should be mitigated as a high priority whilst the lowest scores may have no mitigating actions assigned.

Mitigation plan

For each selected area the actions which are to be undertaken to mitigate the risk should be described. These actions may be targeted at reducing impact, probability of occurrence, difficulty of detection or sustainability threat or a combination of all or some of them.

To check that the mitigating actions are working the risk assessment should be done at regular intervals during project delivery. If the mitigation plans are effective then the risk priority numbers should decrease. When this occurs the focus of mitigation may move to some of the medium priority risks; recognizing that no organization, or Project Team, has limitless resources to address every risk. It is usual for mitigation activities to introduce further sustainability checks or business change planning and tracking activities.

Tool: Stakeholder Engagement Tracker

The aim of this tool (Table 3-14) is to maintain a focus on the status of the stakeholders and their view of, and engagement, in the project. It should track the goals as set out in the Stakeholder Management Plan and give a good overall view of the energy for change.

Table 3-14 The Stakeholder Engagement Tracker explained

Delivery Toolkit – Stakeholder Engagement Tracker					
Project:	*<insert project title>*		**Project Manager:**		*<insert name>*
Date:	*<insert date>*		**Page:**		*1 of 1*
Stakeholder engagement goal					
<insert SMART measures taken from the original Stakeholder Management Plan – usually related to achievement of a specified level of stakeholder engagement and the project vision of success>					
Individual stakeholder engagement analysis					
Stakeholder	**Type**	**Target level**	**Activity**	**Current level**	**Mitigating actions**
<insert the stakeholder name and role>	*<insert type>*	*<insert % or low, medium, high>*	*<insert a description of current management activities>*	*<insert % or low, medium, high>*	*<insert any mitigating actions as a result of the status>*
Summary stakeholder engagement status					
<insert summarizing comments on the effectiveness of the Stakeholder Management Plan>					

Stakeholder engagement goal

The Project Manager needs to ensure that there is a clear goal linked to:

➤ How much the project needs the support of any particular stakeholder or set of stakeholders.
➤ How much the stakeholders need to be actively managed in order to achieve the required level of engagement.

The level of criticality of specific stakeholder engagement to project success would usually be detailed within the Stakeholder Management Plan. For example, at the end of a project there is usually a handover from the project to the business and for this to be successful the business needs to accept it. That acceptance is usually at two levels:

➤ *Level 1* – formal business acceptance
➤ *Level 2* – informal acceptance by people within the business.

Individual stakeholder analysis

Stakeholder

Even if there are distinct sets of stakeholders it is usual to track engagement for each individual stakeholder so that any specific issues can be identified.

Type

This defines the person's role in the project, whether they are sponsoring the project, involved in the project or impacted by the project. Typical categories used during stakeholder management planning are described in Table 3-15.

Table 3-15 Stakeholder types

Stakeholder type	Description	Typical delivery issues
C – Champions	➧ Usually have been involved in initiating the project ➧ They want the project to be delivered and attempt to obtain commitment and resources to do so	➧ Can be used to support areas of high resistance ➧ Need to maintain good communication to keep their engagement
A – Change agents	➧ Usually are integrated into the Project Team ➧ They support the team by taking on business change scope responsibility and maintain an effective link with the project ➧ They have energy for the change and convey this to their colleagues impacted by the change	➧ Can be used to support areas of high resistance ➧ Need to integrate them into the team and use them to channel communications to the business at a lower level (those impacted by the change) ➧ Need to listen to their ideas and challenges as they are likely to know a lot about the current business and therefore the practicality of any changes
S(a) – Authorizing sponsor	➧ Ownership of the project ➧ Accountable for the realization of the business benefits	➧ Can be used to provide a strong link to senior business stakeholders ➧ Need to maintain good communication to keep their active involvement ➧ May need to use to obtain additional resources or support high level change resistance
S(r) – Reinforcing sponsor	➧ Local ownership of the project as delegated by the authorizing sponsor	➧ Likely to be a senior person within one part of the organization impacted by the change ➧ May be given specific authority over a set of resources or a specific role linking to the business ➧ Need to keep them highly engaged
T – targets	➧ Those impacted by the project ➧ These may need to work differently as a result of the project, learning new skills or following new procedures	➧ Anybody in the organization who is impacted by the project and associated business change is a target. The larger and broader the project, the greater the number of targets ➧ Need to find any pockets of resistance ➧ Need to find highly engaged targets and use them to break down pockets of resistance or to improve communication ➧ Need to use the knowledge that targets have and find a way to get participation and therefore energy for the change

Target level

Measuring engagement is notoriously difficult as it aims to measure an intangible. We can usually make reasonable assumptions of engagement through monitoring behaviours, language and actions. For example:

➧ Assign a percentage engagement score based on a combination of observed behaviours and activities/actions related to the project.

- Develop a spider chart profile against key parameters such as alignment of verbal and body language, response to project requests, or meeting agreed commitments. This type of chart needs to have an objective method to determine the score for each parameter.
- Give each stakeholder a RAG score (Red, Amber or Green status) – bearing in mind that not all stakeholders need to be green (fully engaged). This type of scoring tends to be too subjective for detailed analysis but can be used to summarize the status based on other measures.
- Track responses to a survey conducted at key points in the project. This is useful if the questions are posed in the right way to generate a real response. Survey design can take quite some time and requires a good understanding of what the stakeholders care about.

Targets can change as the project is delivered as people move role or as specific parts of the project become more or less important.

Activity

This is the specific activity which will be done to maintain or increase stakeholder engagement. This may include additional communications, face to face meetings, requesting active involvement or working more closely to ensure that the stakeholder delivers what is needed to support sustainable change. The most important stakeholder management activities are listening and watching:

- Listen to what the stakeholder is saying to you about the project, listen to the way it is said – verbal and body language.
- Listen to the way the stakeholder talks about the project when he is not talking to you – does it align? For example you may be told by a stakeholder that he is fully supportive of the project and then hear him telling a colleague that there is no way he's letting you have those resources.
- Watch for aligning behaviours. If a stakeholder is supportive and then follows through with an agreed action in a supportive way then everything aligns. However if a stakeholder agrees to an action and then does not complete it, no matter how often he is reminded, then there are clearly issues.

Current level

This should be measured in the same way and with the same units as the target level of engagement.

Mitigating actions

When there is a negative gap between the target and current engagement levels a mitigation action needs to be put in place to remove the gap on the basis that the target value is still relevant.

When there is a positive gap a Project Manager should consider if the resources to manage this stakeholder could be better used elsewhere whilst not impacting stakeholder engagement.

Summary stakeholder engagement status

At the completion of the Stakeholder Tracker the Project Manager should be reflecting on the cumulative impact of the current level of stakeholder engagement. In particular any issue which is critical in terms of project success should be highlighted and mitigated. Often it is the cumulative impact of a number of small issues rather than one major stakeholder issue which can adversely impact a project.

Tool: Business Change Mitigation Matrix

The risk decision flowchart approach (Figure 3-14) has been adapted for a business change scenario within a project to show the Project Team that sometimes when a business change risk cannot be effectively mitigated it is better to find another way to solve the issue. This is about resource prioritization in an area of the project that typically is resource constrained.

There are four possible outcomes from a risk mitigation review:

- Do nothing – there is no risk.
- Perform a mitigating activity – to reduce the risk impact or likelihood.
- Do something different – solve the required business change outcome a different way because the risks of doing it the current way are too high.
- Stop the activity – the business change outcome is no longer needed.

Figure 3-14 has been converted into a table for ease of documentation of the business change risk mitigation review (Table 3-16). The aim of this tool is to prioritize resources so that they are working on

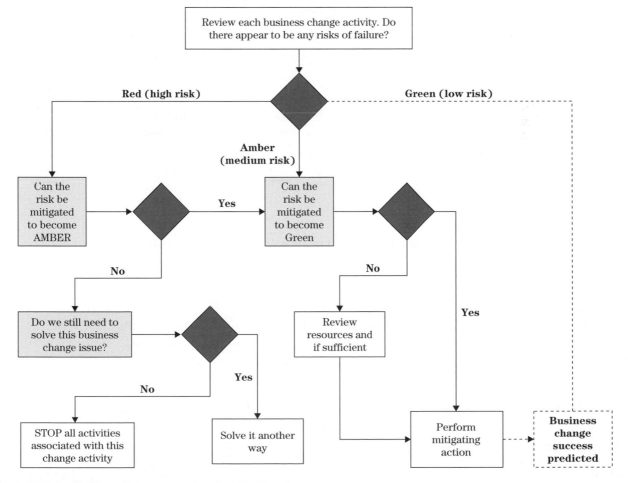

Figure 3-14 Business change mitigation decision flowchart

Table 3-16 The Business Change Mitigation Matrix explained

Delivery Toolkit – Business Change Mitigation Matrix					
Project: <insert project title>			**Project Manager:** <insert name>		
Date: <insert date>			**Page:** 1 of 1		
Mitigation					
Business change activity	**Risk level**	**Risk mitigation possible?**	**Business change outcome needed?**	**Modification possible?**	**Mitigation review outcome**
<insert activity>	<insert potential for failure in terms of red, amber or green>	<insert whether the risk can be mitigated to amber or green>	<insert whether there is still a need for the business change outcome>	<insert if the activity can be modified to achieve the same goal>	<insert the agreed action>
Summary					
<insert a summary of the level of risk in delivering the business changes and the level of resources needed to deliver the business changes successfully>					

the most critical areas of the business change scope, but only those areas that are realistically able to be delivered. In other words, there is no value in spending time and resource changing the business in a way which the business cannot operate and/or sustain.

Mitigation

Business change activity

Each distinct area of business change activity should be listed, whether this is as simple as training or as complex as organizational change or significant changes in ways of working.

Risk level

For each activity ask the first question from the flowchart: 'are there any risks of failure?'. A RAG analysis is used where red is a high risk, amber is medium risk and green is low risk.

Risk mitigation possible?

This column should be filled in as follows:

- *Green risks* – no response necessary – put 'n/a'.
- Amber and red risks-respond 'yes' or 'no'.

Risk likelihood prevention?

Respond 'yes' or 'no' depending on whether the business still needs the outcome of the business change activity. For example, a facility may not need to operate for 24 hours if the main project has been able to increase facility efficiency. Therefore the business change activity to go to a shift system is no longer needed.

Modification possible?

Respond 'yes' or 'no' depending on whether the required business change outcome can be solved by another route. For example, instead of moving current resources on to a shift pattern to extend total production hours the company could just employ a dedicated night shift.

Mitigation review outcome

Detail the outcome of the review (one of the four options). Outline specifically what actions are to be completed, if any.

Summary

At the completion of the mitigation review the Project Manager should be reflecting on the cumulative impact of the current level of business change risk mitigation. In particular the required level of Project Team or business team resources needed to complete the activities and/or mitigate the risks.

Business plan delivery case study – production capacity improvement

To illustrate the key points from this chapter the delivery phase of a manufacturing improvement project is described. It focuses on the delivery of the business plan only.

Situation

A pharmaceutical device manufacturing facility was facing a crisis:

- Reject rates were increasing.
- Work in progress (WIP) inventory appeared completely out of control.
- There were frequent production line stoppages due to a lack of available material.
- An older production line appeared to be operating at better yields than the two newer ones which had never achieved stated design capacity.
- Production lead times were growing and customers were receiving orders later and later.

In addition to this, current sales were forecast to double in the next 6 months as the device is launched into new markets.

Initial project approval

The reaction from the business was to immediately commence a project to increase manufacturing capacity through the addition of more production equipment and resources. A business case was developed to support this (Table 3-17).

Table 3-17 Approved business case – capacity improvement project

BUSINESS CASE			
Capacity Improvement Project			
Business case developed by:	Director of Device Manufacturing Strategy	Date:	January 1st
Project reference number	PR_059	Business area	Manufacturing Strategy
Project Manager	To be confirmed	Project sponsor	Director of Device Manufacturing Strategy
Business background	Within 3 to 6 months the sales forecast for the device will double from 90 packs per month to 180 packs per month (each pack contains 500 devices)Current output matches demand through the use of one older production line (line 2) and two newer production lines (lines 1 and 3)		
Project description	The installation of an additional production line (design capacity of 100 packs/month)		
Delivery analysis	Use of site based project engineers to procure the production line (preferred supplier) and install in line with all appropriate regulations		
Business change analysis	The project is required to support the launch of the device into new markets in line with the business strategy for this product line		

(Continued)

Table 3-17 (Continued)

BUSINESS CASE			
Capacity Improvement Project			
Value add analysis	Cost of investment is $1million (25% accuracy) with a 4 month lead-time for the production line delivery to site. Revenue investment is also required in terms of additional operations staff and other site infrastructure costs. Anticipated returns are additional sales of 90 packs/month worth an estimated $250,000		
Impact of NOT doing the project	If the additional capacity is not delivered then the new markets cannot be supplied		
Project approved (*Value Add or Not?*)	YES	**Name of approver and date**	Vice President Manufacturing

Project Business Plan development

Following approval of the project an experienced Project Manager was appointed from within the site engineering team. She reviewed the business case and established that the project goal was to '*make additional production capacity available to meet market needs*'. On this basis she began the process of project delivery planning commencing with the business plan.

Sponsorship

The Project Manager quickly established a working relationship with the sponsor, the Director of Manufacturing Strategy through the development of a Sponsor Contract and a Communications Plan. The Communications Plan was developed in the knowledge that the business case (Table 3-17) had generated a solution to an identified problem but that all viable options had yet to be considered. The first formal communication was the Project Charter (Table 3-18) and the goal of this communication was to open up stakeholder expectations to a different solution.

Table 3-18 Project Charter – capacity improvement project

Planning Toolkit – Project Charter			
Project: Capacity Improvement Project		**Project Manager:** Jane Jones	
Date: February 8th		**Page:** 1 of 1	
Project description		**Project delivery**	
Sponsor Director of Device Manufacturing Strategy		**Project Team** ➤ There will need to be a core team and then 3 separate kaizen teams ➤ The kaizen teams will be made up of ALL of the line team with each line manager being in the core team ➤ Equipment engineers will also be available ➤ Warehouse manager	
Customer Site Manufacturing Director			
Project aim Make additional production capacity available to meet market needs			

(Continued)

Table 3-18 (Continued)

Planning Toolkit – Project Charter

Project:	Capacity Improvement Project	Project Manager:	Jane Jones
Date:	February 8th	Page:	1 of 1

Project description	Project delivery
Project objectives	**Additional resources**

Project description

Project objectives
- Observe current operation of the 3 production lines and collect performance data
- Conduct 3 kaizen workshop events to design and implement appropriate change on each line
- Modify equipment (as needed)
- Develop standard operating procedures (SOPs) to support sustainability of changes (including performance measures)

Benefits
- Capacity increased on each line so that total capacity meets forecast (180 packs/month)
- Minimal operational running cost increase
- Improved product quality
- Improved process cycle time and therefore ability to meet customer lead times

Final deliverable
- Three operational lines of a specified capacity
- New SOP and performance measures board

Interim deliverables
- Process observation check sheets
- Kaizen workshop design
- Line teams trained in kaizen tools and methodology
- Kaizen output × 3
- Equipment layout modifications
- Equipment modifications
- Pilot SOP and performance measures

Critical milestones Vs deliverables
- End January – all process data
- March 1st – line 1 kaizen (pilot)
- April 15th – line 2 kaizen
- May 5th – line 3 kaizen
- June 1st – handover to BAU

Project delivery

Additional resources
- An external consultant is needed to support the process improvement workshops (kaizen) using the lean six sigma methodology
- A revenue budget of approximately $100k
- Each line will need to be 'off line' for the kaizen week and potentially for the following week to complete modifications

Critical success factors
- Engagement of the line managers and operators
- Accurate data on current operations
- Understanding the complete supply chain process: from customer order, to despatch, to customer
- Ability to modify operation of the line, physical layout of the line and the way that line teams work
- Management support and sponsorship
- Access to lean six sigma expertize

Risk profile
- Root cause for poor line performance may not be easily solved
- Line teams may be defensive and not want to support the changes
- Other parts of the supply chain may not support the changes

Organizational dependencies
- Customer orders – volume and timing
- Warehouse capacity

Project delivery approach
The strategy for this project is to pilot an improvement process on one line and then to follow on and use it on the other lines
The process being followed assumes that the root cause problem can be solved by process changes and minor equipment modifications. This is on the basis that the lines are operating at less than 50% of their design capacity

Benefits Management

In attempting to understand the issue to be resolved, the Project Manager facilitated a benefits mapping session with a group of senior stakeholders to identify the high priority benefits:

- **Ability to meet customer orders on time** (with implied capacity and supply chain speed requirements).
- **Product quality right first time** (thus reducing the cost of poor quality).
- **Profitability of the manufacturing operation** (so that the department would remain a viable part of the long-term future of the organization).

As a result of this session stakeholders were open to three options for delivering the required set of benefits:

- *Option 1* – leave current production lines as they are and buy new a new line.
- *Option 2* – Improve current production lines and buy a new line.
- *Option 3* – Improve current production lines.

Based on a structured benefits-risk analysis option 3 was selected.

Business Change Management

A review of the current situation had already demonstrated that none of the three production lines was operating at or near their design capacity and that if they did there would be no need for any additional equipment or personnel. However, taking a production process from a yield of 46 to a yield of 92% was seen as an impossible goal. Convincing senior stakeholders and then convincing those who are at the heart of the change process are two different things. The Project Manager recognized that formal approval was only the start of 'unfreezing' organizational mindsets about how this particular manufacturing operation was run.

The business change approach was to work closely with the operations team and the senior stakeholder group so that resistance was eliminated and confidence in a positive project was increased. As a part of this the Project Manager had to provide evidence that the project approach was highly likely to deliver the required benefits, which she did based on process improvements at a similar manufacturing plant in the UK.

Approval for delivery

Based on this review of the ability to deliver the highest priority benefits for the business, all stakeholders agreed that Option 3 should be progressed. The draft Project Charter was agreed and approved by all those with high authority and influence and the revised business case was developed and approved (Figure 3-15).

The benefits specification and overall inclusive project approach convinced the senior stakeholders that the project was viable, although not completely risk free.

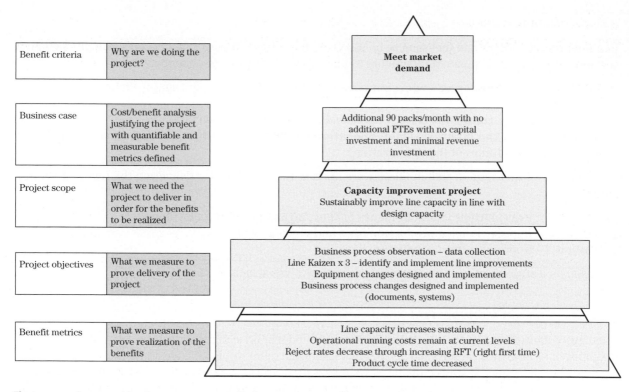

Figure 3-15 Approved business case – capacity improvement project

Business plan delivery

The project commenced with a launch event which included all team members, end users and senior stakeholders. The project roadmap allowed for a data collection phase, a kaizen phase (design and implementation of change) and a consolidation and handover phase. As a part of the go/no go stage gate at the end of the data collection phase the project business plan delivery was reviewed (Table 3-19).

The review highlighted three areas where an increased focus was needed:

- **Communication** – the delivery of the Communications Plan needs to be checked and modified as necessary. This is fundamental to getting senior business support and also the start of engaging the production teams.
- **Stakeholder management** – in order to successfully design and implement new ways of working the production teams need to be engaged. They are not. The risk this lack of engagement poses to the sustainability of change needs to be assessed so that appropriate and timely actions can be taken.
- **Business change management** – the business changes need to be the right ones, so before the design proceeds too far the business change goals need to be reviewed and appropriate actions taken.

The Project Manager and sponsor agreed that if these three areas could be quickly brought back into control the kaizen events could proceed.

Table 3-19 Business Plan Review Checklist – capacity improvement project

Delivery Toolkit – Business Plan Review Checklist			
Project:	Capacity Improvement project	**Project Manager:**	Jane Jones
Date:	Week 6	**Page:**	1 of 1
Sponsor identification			

Sponsor Name
Peter Smith

Sponsor Role
Director of Device Manufacturing Strategy

Sponsor contract review

Is the sponsor operating to the agreed contract?
No – the communication of why we need to collect the data was not consistently communicated to the production teams and as such there was some resistance and suspicion when the team started to observe current ways of working. Having production representatives in the team was not as helpful as we expected due to this

Is the Project Manager operating to the agreed contract?
Yes – but the desire for additional communication is becoming clear. This project is so fundamental to Peter's business accountabilities that he needs to be kept updated more frequently than originally thought

Does the contract need to be amended in any way?
Yes – Need to discuss communications both in terms of how Peter gets data from Jane but also which information needs to be cascaded within the business to support the business change

Business strategy management

Will the business case be successfully delivered?
Yes – at this stage there are growing business change risks BUT it is early enough in the project that their mitigation is possible

Are all stakeholders appropriately engaged?
No – need to engage with the production teams

Is the business ready for the project?
No – not yet but it will be – more work to be done in this area

Summary and action plans

Is the progress of each area of the project business plan according to plan?
No – communications and business change design and management is slightly higher risk than expected at this stage and both areas will need a closer review in the next week or two to ensure that appropriate mitigating actions are in place before the start of the kaizen event (where the actual design and testing of that design will occur)

Summarize any actions needed to support successful business plan delivery
It was always known that business plan delivery success was based on changing the views of the production teams. They viewed an improvement project as a direct comment on their ability to deliver product to the customer on time and within specification. In addition they have had to deal with older equipment and little maintenance support for years and see new equipment as the only solution – and one which they deserve!

Communication

Communication is seen as an essential process when delivering a project which will change the ways of working of so many people and teams in the organization. The fact that the sponsor has not been 100% supportive is a surprise to Jane, the Project Manager. To support some re-contracting with him, Jane reviewed the status of the delivery of the Communications Plan by developing the first revision of the Communications Tracker (Table 3-20).

Table 3-20 Communications Tracker – capacity improvement project

Delivery Toolkit – Communications Tracker				
Project: Capacity Improvement project			**Project Manager:** Jane Jones	
Date: Week 6			**Page:** 1 of 1	
Communications goal				
The communications goal is to engage stakeholder support so that the RIGHT project can be delivered: one which delivers the required benefits at the optimum organizational cost. KPIs being tracked are: ⇒ Stakeholders with high power make timely decisions which support project progress ⇒ Stakeholders with high influence are overtly supportive of project plans				
Key Message	**Audience**	**Activity**	**Feedback**	**Mitigating actions**
1. This project is about improving what we have	Senior Stakeholders	Project Charter verbally reviewed in 1–1's	Sponsor issued the Project Charter by email and most senior stakeholders say they have not have time to read it	Sponsor to schedule 1–1's with senior stakeholders
2. This project is about improving what we have	Production teams	Project Charter verbally reviewed in 1–1's with production team leaders for them to cascade	Team leaders were OK with the level of dialogue with the Project Manager but felt uncomfortable cascading this further	Project Manager and Project Team to take an active role in future production team meetings
3. Status – data collection has started	Production team leaders	1–1 meeting every 2 weeks	Team leaders are using the 1–1's to deliver the feedback from production teams – resistance to letting anyone 'check up' on them	Keep these meetings as a measure of the change of tone as the above occurs
4. Update – 1st kaizen event due to start	All	Monthly 1 page bulletin	Survey before project start and after 1st bulletin issue in week 4 shows no increase in engagement	Reschedule the kaizen team training to support the production teams getting involved with data collection and kaizen planning
5. Kaizen events will take up production time	Senior Stakeholders	Kaizen success criteria and decision via 1–1's	Most stakeholders rejected the meeting requests	Following the initial 1–1 to outline the project goals the sponsor will highlight the key decision regarding resources and a go/no go decision
Communications summary				
The communications seem too formal and distant considering the huge impact the project will have on production teams. Suggest that in future the Project Team and Project Manager go to production team meetings to be the 'face' of the project				

   The original intent was that the Production Team Leaders were best placed to pacify and cajole the production teams who were very vocal and sceptical about any form of improvement project. Clearly this approach has not generated the quick turnaround needed for success, so a more involved, fast track way of getting production teams engaged is to be tried over the coming weeks (as the first kaizen event is scheduled for week 10, only 4 weeks away).

Stakeholder Management

The current lack of engagement by the production teams was reviewed using the sustainability FMEA (Table 3-21) based on assumptions of business changes as a result of the three separate kaizen events, one per production line.

Table 3-21 Sustainability FMEA – capacity improvement project

Delivery Toolkit – Sustainability FMEA						
Project:	Capacity Improvement project			**Project Manager:**		Jane Jones
Date:	Week 6			**Page:**		1 of 1
Business change area	**Probability of not**	**Impact of not**	**Ability to detect**	**Sustainability threat**	**Risk priority**	**Mitigation plan**
Team training on continuous improvement techniques	3	4	1	5	60	Complete training before the kaizens commence
Bring testing into the main production unit	4	5	3	3	180	Ensure that testing staff are involved in the kaizens
Change team working to cross functional/cell based	5	5	4	4	400	This is about engaging the team in the possibility of lean solutions – these may not be the exact solutions but unblocking material flow is likely to involve some form of value stream approach. Customize training to introduce these design scenarios.
Use kanbans rather than scheduling	5	4	5	5	500	
Stop production according to kanbans	5	5	5	5	625	
Revised organizational structure	3	4	5	1	60	Need to commence organizational design or at least be open to this as a solution from the kaizen workshops

(Continued)

Table 3-21 (Continued)

Delivery Toolkit – Sustainability FMEA						
Project:	Capacity Improvement project			**Project Manager:**		Jane Jones
Date:	Week 6			**Page:**		1 of 2
Business change area	**Probability of not**	**Impact of not**	**Ability to detect**	**Sustainability threat**	**Risk priority**	**Mitigation plan**
Change to working hours	5	3	3	3	135	If this solution is not highly likely then it needs to be rejected as early as possible
Scoring system						
Probability	1 = low 5 = high	Impact	1 = low 5 = high	Detection	1 = early 5 = late	Threat	1 = low 5 = high

The assumptions were based on similar improvement projects done on production lines with similar issues and the FMEA was useful in testing potential design scenarios. The Project Manager used this analysis to customize the training programme and start to build energy for lean solutions within the production teams. By involving the team in all aspects of the design and then implementation via a 'blitz' approach the Project Manager was hoping to get a sustainable solution with minimal overt resistance.

However the analysis also uncovered the risks of not informing senior management of the likely changes ahead. They were used to seeing a busy production area piled high with inventory and everyone rushing around sorting out problems. They needed to be engaged in the potential new vision of the production area:

- Inventory in designated areas used as a visual signal to either start or stop a particular part of the production process
- Some production areas not operational (because there was sufficient inventory for the next manufacturing stage already)
- Teams or cells working more effectively together mitigating problems rather than fighting fires.

In other words their paradigm for a production environment had to be completely rewritten. This was a stakeholder issue that the Project Manager had not anticipated. As a result a training session was held just for senior stakeholders. This focused on the practical demonstration of a lean process.

Business change management

As a final part of the review, and following on from the sustainability FMEA, Jane conducted a different type of risk assessment to check that the there are no better ways to solve the business improvement goals than to take the designed and planned change approach (Table 3-22).

Table 3-22 Business Change Mitigation Matrix – capacity improvement project

Delivery Toolkit – Business Change Mitigation Matrix					
Project:	Capacity Improvement project			**Project Manager:**	Jane Jones
Date:	Week 6			**Page:**	1 of 1
Mitigation					
Business change activity	**Risk level**	**Risk mitigation possible?**	**Business change outcome needed?**	**Modification possible?**	**Mitigation review outcome**
Collection of current performance data	Red	No Too much resistance in production teams	Yes Current performance is needed in order to start root cause analysis	Yes Data can be collected in different ways	Use the production teams to collect the data themselves following their training
Continuous improvement training programme	Amber	Yes Can face resistance head on	Yes Production teams need to be skilled in this area if the change is going to be successful	Yes Training can be adapted to suit a very resistant audience	Training can be in production teams and very practical to convince the teams that these solutions do work
Conduct 3 × production line kaizens	Red	Yes Can face resistance head on	Yes Need to improve production	Yes But the timescale for alternatives does not align with business needs	Do a pilot kaizen with the most engaged production team and invite other production teams to observe
Conduct an organizational design workshop	Amber	Yes	Unsure	Yes	HOLD on this activity until the pilot kaizen is completed
Summary					
Although this is a relatively high risk project there is potential to mitigate risks through a more stakeholder management intensive approach. This will take a lot more resources but ultimately ensure that the changes are delivered within the required timescale					

This type of risk analysis demonstrated the huge impact of the resistance to change within the production teams. Considering that they form the bulk of the Project Team their engagement is fundamental. Therefore in order to get them fully engaged the Project Manager had to understand their resistance and develop strategies to remove it. Jane got the teams together during one of the breaks (when the lines were all shut down) and asked them about the ways things were at the moment. Based on the analysis of resistance she was able to strengthen the drivers for change by involving the production team members at each stage and getting them to fully own solutions they had thought of. The training, communication and kaizen events were all designed to prove that the production area could work in a different way.

All other mitigation plans in the various risk analyses were delivered with the resulting success during the pilot kaizen, quickly followed by the other two.

Conclusions

As a result of the Business Plan review during week 6 a series of risk mitigation activities were completed. This resulted in a successful pilot kaizen quickly followed by the two further kaizen events.

At the end of this project capacity was increased without requiring the installation of an additional production line and associated resources to operate and manage it. Each of the existing lines was able to sustainably deliver the required increased capacity which did occur approximately in line with forecasts (Figure 3-16).

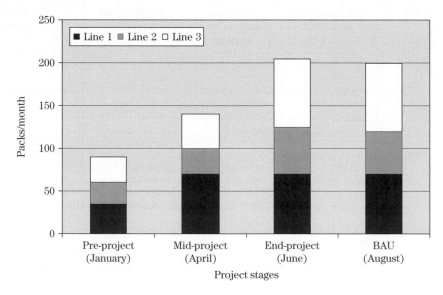

Figure 3-16 Benefits tracking – capacity improvement project

This required a major business change in terms of:

➤ Understanding the business processes required to deliver the product 'right first time'.
➤ Eliminating the root causes of the low yield including the people issues such as functional behaviours.
➤ Operating the supply chain differently to focus on bottleneck management, 'right first time' production and empowered production teams.
➤ Materials management through the use of kanbans which further empowered the cross functional teams, allowing them to determine when production should start or stop based on predetermined levels of inventory at each stage of the manufacturing process.

Key Points

The aim of using this particular case study was to demonstrate:

➤ The criticality of stakeholder management – planning to have engaged stakeholders and then adapting those plans and mitigating stakeholder risks throughout delivery.
➤ Project outcomes, and therefore business outcomes, can be achieved any number of ways – even the most obvious business case can benefit from a review from a new perspective.

There are many ways to deliver a set of benefits. The business plan is simply there to set out the selected way in which these benefits will be delivered. However as with any plan, the business plan needs to be flexible enough to cope with situations which become apparent as the project is delivered.

Troubleshooting business plan delivery

Table 3-23 is a list of common issues associated with business plan delivery.

Table 3-23 Troubleshooting business plan delivery

Typical symptoms	Example root causes	Example solutions
1. Not following the business plan	Poor or no sponsor-Project Manager relationship	Ensure that the project has an appropriate sponsor and that a contract is developed with the Project Manager
2. No apparent link between the project and the business		
3. Poor sponsorship (for example, not achieving the sponsor contract)	Sponsor does not have either the time or the understanding to be active	Ensure that the organization has an understanding of the role and value of a sponsor
	Inappropriate sponsor selected	Identify the organizational position of the selected sponsor and check that they are at the lowest level in the organization with authority over the people/business processes impacted by the project
4. Sponsor changes	Organizational changes	Re-contract with the new sponsor ensuring that the power/influence matrix has been considered. Think about using a reinforcing sponsor to maintain continuity
5. Sponsor does not actually believe in/engage with/act as a sponsor	They have been 'given' a project to sponsor and are not committed to it or interested in it	Identify a reinforcing sponsor who could have delegated authority from the sponsor
	The Project Manager does not have a good relationship with the sponsor	Develop a contract with the sponsor and work to it
6. Sponsor does not have the influence he/she thought	Changing stakeholder environment	Encourage the sponsor to contract with all new stakeholders
7. Sponsor does not have the authority he/she thought	Inappropriate sponsor selection or organizational changes	Select new sponsor and check that they are at the lowest level in the organization with authority over the people/business processes impacted by the project
8. Poor or no stakeholder engagement	No Stakeholder Management Plan and/or not delivering the Stakeholder Management Plan	Develop a Stakeholder Management Plan, enact the plan and measure level of stakeholder engagement frequently
	No structured communications to stakeholders	Develop a Communications Plan so that appropriate messages are delivered to each type of stakeholder

(Continued)

Table 3-23 (Continued)

Typical symptoms	Example root causes	Example solutions
9. Inappropriate communications (wrong messages) 10. Ineffective communications (wrong vehicle)	Inappropriate communications planning or not delivering the Communications Plan	Align the Communications Plan with the Stakeholder Management Plan to ensure the right messages are delivered in the right way. Make sure that the Communications Plan is amended if it's not working
11. Project does not look like it will deliver the intended benefits	No Benefits Realization Plan	Develop a Benefits Realization Plan and get this approved by the sponsor
	Market conditions have changed – the benefits cannot be realized Benefits assumptions were wrong to start with Timescales have changed – benefits will not be achieved at the predicted time	Review the Benefits Realization Plan with the sponsor and together challenge the project – make changes as necessary, for example: ➤ Change the project (scope, timescale) ➤ Stop the project ➤ Review the business case (change business expectations)
	Benefits not capable of being realized	Stop the project and conduct a full review – is this the right project to be progressing?
12. Business states that the benefits are no longer appropriate?	The business case is no longer valid	Stop the project and conduct a full review – is this the right project to be progressing? Develop a new business case
13. Business is disengaged from the project	No Business Change Plan or poor definition of the changes needed by the business	Develop a Business Change Plan with the business owners Ensure that Stakeholder Management Plan includes all those whose BAU will need to change as a result of the project
14. Projects have conflicting requirements for business changes (for example, resources, timing, how the business will operate)	Weak link to other projects in a programme of change	Develop a Business Change Plan with the business owners ensuring that all business changes are aligned (both within and outside of the project or programme)
15. Current business is adversely impacted by the project	Poor management of business risks (to do with the management of the business environment)	Ensure that the Business Change Plan considers how to get from current to future state without adversely impacting the business

(Continued)

Table 3-23 (Continued)

Typical symptoms	Example root causes	Example solutions
16. Resistance to change in the business	Fear, uncertainty, doubt – what will the project do to me/my job?	Good communications in order to engage all those impacted by the changes. Consider other methods of reducing change resistance such as active participation if appropriate
	Technical change, not business change (the initiators of the change are OK – everyone else is left to follow)	Develop a Business Change Plan to ensure that all other parts of the business impacted by the change are aligned
	Lack of communication – no one knows what's going on	Ensure that there is good stakeholder communication in order to ensure engagement with all those impacted by the changes
	Business does not believe that the project is valid	Review with sponsor and consider other ways to engage stakeholders
		Consider if the view has any validity (maybe they know something you do not)
17. No implementation of business changes	No allocated resources in the business to make these changes	Develop a Business Change Plan and get buy-in from all appropriate stakeholders
	Changes might have a lower priority within BAU than in the project	Ensure that the plan is time-based and fully resourced (people, funds and assets)
	Business did not know it had the changes to make	
18. Organizational tendency to 'reset' to original state – lack of sustainability	No Sustainability Plan	Develop a Sustainability Plan and get buy-in from appropriate stakeholders including business owner
	No buy-in to the required business changes	Ensure that the sponsor has contracted with the business owner so that each are aware of what needs to occur to achieve success
	Benefits of changes have not been seen – too early	Ensure that all stakeholders understand when benefits will be realized – good communications
	Benefits of changes have not been seen – not delivered	Conduct a project review to identify why benefits have not been realized

Handy hints

Business plan delivery is about managing and maintaining good relationships between the project and the business

It is easy to forget that this element of the project is about relationships and not just the delivery of a business case. Project delivery management must effectively integrate both the 'hard' and 'soft' elements if it is to be robust and effective. The Project Manager and sponsor should be using the business plan as a working document which supports their relationship as well as a robust link to the business.

Be able to summarize the key messages from the business plan to support delivery of the overall project

It is important that you (the Project Manager) and the sponsor communicate consistent messages to each stakeholder group.

Sponsors need guidance and management

The majority of senior management who are assigned as sponsors will have had a varied experience of sponsoring projects. A good Project Manager will recognize the importance of guiding the sponsor and being very clear about what is needed from him, managing the delivery of the contract between them.

Having an active sponsor takes work

The sponsor, the business and the Project Manager all have to work together to ensure that the sponsor participates in an active and engaged way.

Customers need management

Behave as a consultant to the business when delivering the business plan – it needs these skills to effectively manage the diverse range of stakeholders. Remember that the partnership of Project Manager and sponsor need to effectively manage customer 'needs' and customer 'wants'.

Remember that the approved business case is your contract with the business and the business plan is your contract with the sponsor

Treat it as a baseline to be managed.

And finally...

Effective business plan delivery:

- Maintains a 'live' link between the project and the business – making sure that the project being delivered is needed by the organization.
- Ensures that the set-up plan delivery is appropriate and capable of supporting project success.
- Ensures that the control plan is appropriate and capable of supporting project success.

4 Set-up plan delivery

In the context of project delivery, the project set-up plan is the way we administer the project. It is based on the approved project business case which explains 'what' the project needs to deliver to meet the business needs. The robust delivery of the set-up plan relies on the Project Manager and Project Team working together to ensure that the project is delivered in a value-add way. The two concepts which are fundamental to set-up plan delivery are:

➡ *People management* – the continued engagement of the Project Team and management at a macro and micro level.
➡ *Value management* – the effective management of value so that the project delivers a lean solution for the business.

In order to manage these concepts, the Project Manager needs to manage specific stakeholders effectively (team members and customers) and use Lean Six Sigma tools to support that management.

What is a project set-up plan?

As defined in detail in *Real Project Planning* (Melton, 2008), a project set-up plan is that part of a Project Delivery Plan (PDP) that is the formal articulation of HOW the project will be administered. The goal is to assure the business of the certainty of outcome with respect to the original business case. It covers the following four planning and delivery themes:

➡ Project organization.
➡ Project type.
➡ Project scope.
➡ Project funding strategy.

How to manage delivery of a project set-up plan

To track the delivery of the project set-up plan, a Project Manager must behave both as a Lean Six Sigma consultant and as a people manager in order to actively manage the set-up strategy (Figure 4-1).

The Project Manager has to consider how best to manage the agreed set-up strategy for that project, considering all 'hard' and 'soft' aspects. Although the key relationship in delivering the set-up plan is with the Project Team, the Project Manager also needs to consider the wider stakeholder group, particularly the customer and end-users who can support value management in particular.

The Project Manager as a people manager

The people management skills required of a Project Manager are often underrated. However, without them the management of the Project Team and the customer or end-user group is impossible.

Managing the Project Team

The Project Manager needs to behave as the line manager for the people in his team even though this is a temporary situation, and it is likely that each team member has a 'home base' in the organization with an associated full-time line manager. The role of temporary line manager means:

➠ Getting to know each team member – understanding their strengths and weaknesses and their role in the success of the project.
➠ Finding opportunities to further develop each team member.
➠ Building a motivated group of individuals (a team) so that their combined effort is greater than the sum of the parts (individual team members).

Managing the customer or end-user group

The Project Manager needs to behave as a champion of change for the people in the customer or end-user group. After all it is these people whose working lives will be changed as a result of the project. The role of a champion means:

➠ Getting to know key team members in the customer group – finding those who can become change agents by getting the remainder of the customer group motivated by the changes.
➠ Finding opportunities to engage and involve these key team members.
➠ Energizing a customer team so that they are welcoming of the change and actively support it.

The Project Manager as a lean manager

There is only one question a lean manager will ask as the project progresses – 'does this add value?' This question may be linked to the scope (tangible value delivered as the output from the project) or the project process (intangible value generated as a part of project delivery). As stated in *Real Project Planning* (Melton, 2008) – value is a 'feature, condition, service or product that the customer considers desirable, and which is delivered to them when and where they want it' (Adams et al., 2004). In other words, in a project context, value is a function of scope and project process.

Lean project scope

Project scope is only of value if it is required to enable the defined and agreed business benefits. Therefore the role of the Project Manager in managing value in this context is clear – if scope isn't linked to benefits, then it should be removed.

Lean project process

Project process is only of value if it moves the delivery of tangible project value (scope) forward. Therefore the role of the Project Manager becomes that of 'waste manager' (Figure 4-2), identifying and eliminating any waste from the project delivery process. A lean project is one where each stage,

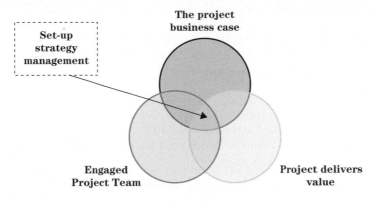

Figure 4-1 Set-up strategy management

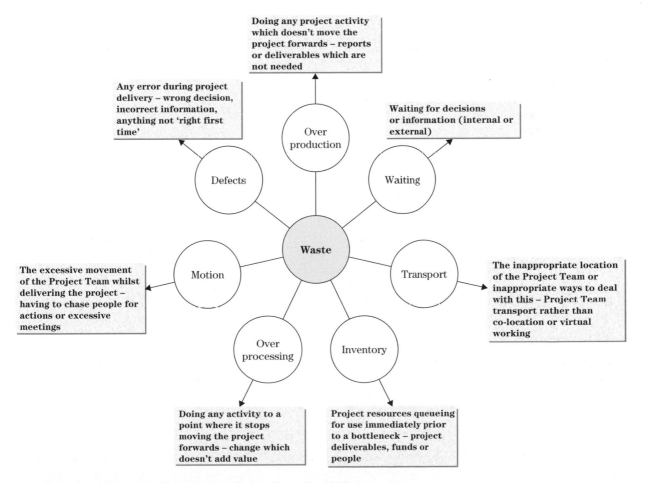

Figure 4-2 The seven types of waste as applied to the project delivery process

activity or decision in the delivery process contributes to the tangible delivery of value and therefore the enabling of the required business benefits.

Lean thinking starts with the customer and the definition of value. Therefore, as a project is a vehicle to deliver value to a customer, the principles of lean thinking should be applicable to project delivery and specific project management processes within that.

A project which is not lean:

➡ Is not necessary.
➡ Is completed late, over budget or in non-compliance to scope objectives.
➡ Is not able to realize the business benefits for which it was authorized.

Waste can be removed from many steps in the project delivery process, from how stakeholders are managed to how the final tangible project deliverable is developed. However, to be truly lean, all these elements need to be linked within a robust business process – to ensure the flow of value as delivery progresses. The lean Project Manager uses Lean Six Sigma tools, techniques and processes to deliver value (Adams et al., 2004).

Project organization

During the project planning stage, considerable effort will have been used to select and develop the Project Team so that they can work effectively during project delivery. Therefore the goal during project delivery is to appropriately manage the team so that they continue to perform and support project success. The focus is on managing people and to an extent using the project as an opportunity to develop people.

Team performance management

There are three areas of team performance which need to be effectively managed during project delivery:

➤ **Activity management** – ensuring that the team do the things they need to do to deliver the project.
➤ **Capability management** – ensuring that the team either have or develop the skills to deliver the project.
➤ **Relationship management** – ensuring that the relationships within and external to the team are supportive and aligned to achieving the project goals.

All three require an element of team administration support as well as active management by the Project Manager. For example, in order to:

➤ Deliver completed project activities the team need tracking tools, access to project documentation, regular progress updates, and a place to work.
➤ Develop enhanced skills the team need training and training resources.
➤ Maintain good working relationships the team need regular interaction and contact, through project team meetings and informal get-togethers.

However, all are reliant on building a customized team (the ideal state) rather than being given a team, a team structure and/or specific team members (usually less than ideal).

Activity management

A performing team will carry out the project activities in line with the plan so that the scope quantity, quality and functionality are delivered to enable the required business benefits.

A key part of assessing the performance of the team is in tracking whether each team member is delivering their part of the overall scope. The RACI Chart (Melton, 2007, 2008) should be used as a baseline to check whether all the activities are being completed by the appropriate team members, for example:

R = responsible

Check that the team member allocated to this activity has actually completed the task. They will have been allocated the task due to their particular skill set. There is usually only one person taking full responsibility for the completion of an activity or deliverable. Where a person, such as a team leader, is taking responsibility for a set of activities or deliverables, there would usually be a more detailed RACI chart for this which is managed by the team leader.

A = accountable

Check that the person in overall charge of the activity is happy with the outcome. Depending on the type of activity, this level of accountability may be seen as the approval of the activity outcome such as an interim deliverable. Only one person can be held accountable for an activity.

C = consulted

Check that team consultation and involvement is progressing appropriately. People are consulted for a number of reasons:

➠ Participation during the progress of the activity by a team member with a specific capability.
➠ Technical review by a team member with a specific capability.
➠ Review by a stakeholder who needs to be engaged in the activity as a part of ownership of the outcome or deliverable.

Most activities need some form of support for completion, and the Project Manager should ensure that the specified and required support is available.

I = informed

Check that team communication of the completed activity is effective and accurate. The total number of people who need to be kept informed of the progress or completion of an activity will depend on how the information would be used to progress other project activities (or as a part of the Stakeholder Engagement Tracker or Communications Tracker).

An example of a completed RACI Chart is shown in Table 4-1. This is taken from a small office reorganization project where one team member was asked to manage the project for the team manager (sponsor).

Table 4-1 Example RACI Chart

Project Management Toolkit – RACI Chart					
Project: Office reorganization			**Date:** Week 1		
Names → Activity ↓	Sponsor	Project Manager	Team member A	Team member B	Team member C
Develop new office layout	AI	R	CI	CI	CI
Develop a cost estimate	I	AR	–	–	–
Cost estimate approved	AR	CI	–	–	–
Manage external suppliers	AI	R	C	–	–
Manage office changes	AI	R	CI	I	I

The RACI Chart can be converted into a tracking tool for use by the Project Manager and Project Team (page 95). It is also used as the basis for activity progress measurement, such as the project schedule (Gantt Chart or milestone chart). It is important that activity-tracking tools are overt and available to the team:

- The team should be motivated by the sum of their individual achievements.
- Any area of low performance is obvious to all – thus using an element of peer pressure and support to highlight and mitigate poor performance.

For any Project Manager, the two areas of performance being managed via a review of activity completion are:

- Activity-completion effectiveness – the right things being done by the right person.
- Activity-completion efficiency – the right things being done in the right way at the right time.

Capability management

A team is pulled together to deliver a project, and the sum of each member's capability is the team's capability to deliver the project. Therefore, during project planning, a great deal of time and effort needs to be spent getting and setting up the right team.

Capability profile

A Project Manager needs to understand and manage the capability profile of his team and be aware of any changes to that profile which may be caused by:

- People joining or leaving the team.
- People not performing or not developing as expected.

A capability profile is an effective tool to review team strengths and weaknesses. Once these are known, a Project Manager can choose strategies to strengthen the strengths and eliminate the weaknesses.

Capability risk assessment

During project delivery, the Project Manager needs to manage that combined capability and develop it as appropriate to the project requirements. A key part of that management is the identification and mitigation of risks.

A Project Manager should be reviewing the capability profile as it changes during delivery and review this against identified risks and associated mitigation plans. For example, if a key capability is within only one team member, then a strong mitigation plan is needed to ensure that either the team member remains on the project team for the duration of the project, or a succession plan is in place. The latter is usually appropriate for a long-term project.

Capability development

A Project Manager will have selected team members who are expected to develop specific capabilities as the project progresses and then use those to support delivery of the project.

Individuals and teams need support in order to develop capabilities including:

- *Formal training* – to build knowledge and skill level 'off the job'.
- *On the job training* – to build capability as the project is delivered under the support of another team member.
- *Mentoring and coaching* – to build specific elements of knowledge, skill and/or behavioural development through a defined 1-1 relationship during project delivery.

Short case study

A small Project Team has been formed to manage a sensitive business-change issue within a large event management company: the outsourcing of a major business function (administrative support). The six team members were selected in order to provide a collated team capability profile (Figure 4-3) in order to:

➤ Deliver the most appropriate solution to match the business needs.
➤ Manage the sensitive change situation as staff transition to a new company.

The small team of six senior managers was maintained as a stable team throughout the project and together they were capable of successfully developing and delivering an appropriate, practical solution. The eventual outsourcing of the administrative functions proceeded with minimal short-term negative impact on the business and maximized sustainable long-term benefits (linked to cost and flexibility of approach).

Capability level		A The events management business	B Outsourcing process and projects	C Project management	D Business change management	E Getting things done	F Stakeholder management
	5 A key strength	X		X		X X X	X X
	4 Unconsciously competent	X		X	X X X	X	X X X
	3 Consciously competent	X X	X X X X	X X X		X X	X
	2 Consciously developing	X X	X X	X	X X X		
	1 A known weakness						
	0 None						
Capability		A The events management business	B Outsourcing process and projects	C Project management	D Business change management	E Getting things done	F Stakeholder management
		Knowledge		Skills		Behaviours	

Key: **X** Individual capability assessment   ▢ Team average capability

Figure 4-3 Short case study – capability profile

Relationship management

An important aspect of team management is how people work together as the project is being delivered, (their interpersonal relationships). This can be considered in three ways:

- *Facilitation* – how the Project Manager chooses to interact with the team.
- *Team working* – how the team members interact with each other.
- *Mentoring and coaching* – how team members work within specific 1-1 relationships for development purposes.

Facilitation

Depending on the nature of the relationships, a Project Manager may manage this aspect in a number of ways. The facilitation model for Project Team management (Figure 4-4) describes two interacting criteria:

- *Level of intervention*: This describes the amount of involvement and direct management from the Project Manager, either high or low. An indication of this is the percentage of time the Project Manager spends with his team.
- *Level of innovation*: This describes the different ways with which the Project Manager may interact with his team. The more customized a process, the more innovation has been used to manage the team.

The result of having all possible combinations of these criteria is four different modes of facilitation for team management.

1. *Content manager* – For most Project Managers, this is the normal mode when starting to work with a Project Team. The Project Manager introduces the process, which the group will follow for a particular situation, and is involved in the development of the final outputs.

		Content Manager	**Influencer**
Degree of intervention	High	Typical mode within a newly formed team with relationships in early stages Project Manager likely to use a traditional process and strongly influence direction and content	Typical mode within a highly performing team with no relationship issues Project Manager likely to use novel activities and provide direction rather than instruction
	Low	**Process Manager** Typical mode within a normally performing team with no relationship issues Project Manager likely to use a traditional process and targets and then let the team use it with minimal support	**Games Master** Typical mode within a highly performing team with good interpersonal relationships Project Manager likely to use a novel game/process and allow the team to use it with minimal intervention
		Low	**High**
		Level of innovation	

Figure 4-4 Facilitation modes for team management

For example, during a project meeting to review supplier performance, a Project Manager would outline the supplier review process. He will be highly involved in the meeting and, as the chair, will strongly influence the outcome through direct questioning and responding to questions from team members.

2. *Process manager* – Project Managers recognize that some situations do require less hands-on involvement. This mode can be used when a Project Team needs to be heavily involved but within a fairly standard process where the outputs are typically owned by the Project Team.

For example, during a project risk-review session, the Project Manager would introduce a standard risk management process and leave the team to review risks and associated mitigation plans and to identify new risks, describe the consequences and score in an objective way (the scoring system is usually defined beforehand). Individual team members would own specific mitigation plans and the Project Manager looks at the overall impact of these.

3. *Influencer* – Sometimes a project situation requires some innovation if the appropriate solution is to be identified. In order to do this, the Project Manager needs to rely on the expertise within the team and allow them the freedom to develop the right solution. However, due to the particular situation, the Project Manager still needs to influence the direction (usually linked to business benefits management).

For example, during the delivery of a technical design phase the Project Manager may facilitate a 'brown paper' scheduling exercise. This is a technique where the schedule begins as a large piece of brown paper on a wall. The team members use post it notes to suggest the technical tasks, time-scale and dependencies whilst the Project Manager challenges them.

4 *Games master* – Sometimes a project situation needs a Project Manager to be innovative but less actively involved. This tends to be when the team need to use their 'technical' expertise.

For example, a Project Team designing an improved manufacturing process decide to use a simulation of the improved manufacturing process using building bricks and involving the operators (end users). The Project Manager ensures that the 'game' is set up appropriately and then stands back to watch the game being played. If the designed improvement is robust, then it will be capable of handling the simulated situations and the operators will have been trained in a safe (and memorable) environment.

Team working

Teams require their Project Manager to be more or less involved as described previously. However, at times, their needs may conflict with the Project Manager's views. This can cause conflict between the team and the Project Manager. To mitigate these, typical approaches can be used:

⬤ **Team building** – There needs to be a culture of continual team building. On projects being delivered over a lengthy duration in particular, there can be team members leaving and joining throughout its lifespan (sometimes as a part of the plan), and this needs to be managed. Typically the stage gates in the project roadmap (page 99) can be used to hold team events which can celebrate success, review performance and plan for the next stage.

⬤ **Communication** – Appropriate communication, particularly of changes in management approach, is fundamental to maintaining a good relationship between the team and the Project Manager. An internal Communications Plan should be developed to ensure that change, progress and team achievements are appropriately and frequently communicated and cascaded to all team members.

⬤ **Consistency** – A Project Manager needs to behave in a way which builds trust within the team. In *Real Project Planning* (Melton, 2008), a continuum of empowerment was described. In essence, this model states that trust is gained by defining how the Project Manager will operate with an individual or team and then by operating in that way. For example, if a Project Manager delegates authority to a team member and then makes the final decision himself, he is starting to build a cycle of distrust; the more he contradicts himself, the more the team will not trust him. The converse is, therefore, also true.

In addition, poor interpersonal relationships with the team, ultimately leading to overt conflict, can occur. In this situation, the Project Manager has a clear role – to facilitate resolution for the better good of the project, the Project Team and the individuals involved. Conflict resolution would typically involve some form of mentoring so that the root cause of the conflict is identified, as well as resolved in the current situation. Problems that are not fully resolved will only surface again later in the project.

Good team working will enable the team to do their part in delivering the project, and it is an important aspect of overall team management.

Mentoring and coaching

In specific situations, individual 1-1 mentoring or coaching is necessary. This may be linked to capability, performance, developmental, behavioural or relationship issues and is distinct and different from usual line management goals, although it should align with them.

For any mentoring or coaching scenario, the general process shown in Figure 4-5 can be applied. The critical step in the process is the first – establishing the right person to mentor or coach:

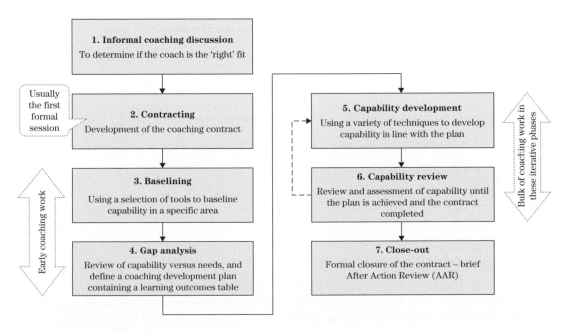

Figure 4-5 Coaching process

➤ Project Managers often get coaching from the sponsor, particularly if working in new area of the business, or an assigned Project Director if the project is particularly large, complex or a part of a bigger programme.

➤ Team leaders within the Project Team can be coached by the Project Manager which is particularly appropriate if the person is looking to gain experience and insights into project management.

➤ Team members within the Project Team can be coached by their Team Leaders in terms of 'technical' expertise, or the Project Manager for project management expertise.

The coaching is made specific to the individual at each stage, although some basic tools are generic so that a clear framework can be established:

➤ *Coaching contract* – a way of contracting so that expectations are clear and agreed (Figure 4-6).

➤ *Learning outcomes table* – a way to measure success during the progress of the coaching or mentoring (Table 4-2).

Scope	Ground rules

• John and Mike will meet 1:1 approximately every 3 weeks at John's scheduling
 	
• The scope of the 1:1's is as per 'John's needs' but in summary is to cover:
   • Development of consultancy capability – specific knowledge skills and behaviours
   • Support on consultancy assignments
 	
• The coaching will continue until the achievement of the learning outcome goals
	• John is to drive the coaching process – the when, what and how
	
	• All coaching sessions are confidential unless explicitly agreed otherwise (2-way)
	
	• Actions/agreements/commitments will be kept by both John and Mike
	
	• John will keep Mike updated on key issues/scenarios as agreed during the sessions

Vision of success	Communication
• Development of John's consultancy capability as measured by the achievement of learning outcome goals • Development of John's consultancy role within the Business Consultancy team • Development of John's confidence and also his reputation (how others see him)	• Input to 1:1 sessions *must* be driven by John – the agenda, the preparation, etc. • John must outline the project scenario (past or present), the issue and proposed solutions • John will track how he responds to certain stimulus (different people, different situations) • In consultancy skills/knowledge development, the sessions will be driven by current assignments

Figure 4-6 Example coaching contract

The mentoring or coaching should continue until the 'coaching contract' is complete and the learning outcomes have been achieved. A key part of meeting these mentoring and coaching expectations is establishing an appropriate 2-way relationship based on:

➤ Mutual trust – by both adhering to the coaching contract.

➤ Openness and honesty – by being frank in both the current situation and feedback. A good coach will be able to read body language to test whether there is any conflict between the verbal and body language.

Table 4-2 Example learning outcomes table

Learning outcome	How achieve outcome	Development outcomes			Comments (Month 7)
		Red	**Amber**	**Green**	
An area of knowledge, skill or behaviour	*Development method*	*Minimal outcome observed*	*Outcome progresses with support*	*Outcome achieved*	*Comments on progress at review point*
Knowledge – Develop a 'user's' understanding of the consultancy process	Work with coach to develop knowledge as consultancy assignments are delivered	Baseline Month 1	Target Month 7 One assignment delivered	Target Month 12 Three assignments delivered	Now assigned to two separate consultancy assignments and using process. One completed and the other 75% completed. Excellent feedback to date.
Skill – consultancy tools development to expert level: assignment roadmap, charter and contract plan	Review current use of these tools with mentor and develop	–	Baseline Month 1	Target Month 7	All three tools used effectively on both current assignments
Skill – consultancy tools development to expert level: benefit maps, measures maps	Attend 'lessons' with mentor and then review the use of tools on current assignments	Baseline Month 1	Target Month 7 for lessons	Target Month 12 for expert level	Not really had had an opportunity to use these much without lots of support
Skill – business improvement tools development to expert level: multiple cause, interrelationship and time value maps	Attend a kaizen event being lead by mentor and get live use in these tools. Use and review on current assignments	Baseline Month 1	Target Month 7 for kaizen	Target Month 12 for expert level	Attended kaizen and took a lead in one of the sessions. Need to find more opportunities for large group facilitation of these tools
Behaviour – to gain confidence as a consultant		Baseline Month 1	Target Month 7	Review Month 12	Feedback from customers on excellent performance on assignments to date. Seeing increased confidence and capability at coaching sessions.
Behaviour – to develop a broader range of facilitation styles		Baseline Month 1	Target Month 7	Review Month 12	Still seeing a high degree of intervention. For example, at the kaizen event John still took a lead around the flip chart – still very traditional

Coaching summary
John has progressed in all areas and in particular is demonstrating an increased level of confidence, capability and hence performance. Continued mentoring in specific areas is required and perhaps assignment to some more complex problems with larger teams and more complex stakeholder issues.

Tool: Tracking RACI

The aim of this tool (Table 4-3) is to ensure that the Project Team is operating as planned during delivery. Although it will check whether an activity has been completed or not, the main goal of this tool is to check if the appropriate team member has completed that activity. In specific circumstances, it is critical that activities, tasks or deliverables follow a specified route from generation, through review and approval to completion and handover. This tool can be used to track only those activities which have this criticality, as a way to manage one element of project risk.

There are other tools that focus on completion of activities in terms of meeting cost, time, quality, quantity and functionality criteria (pages 110 to 117). This tool would generally be used as a part of an internal team review rather than as a progress-tracking mechanism and is demonstrated in the case study at the end of this chapter (page 123).

Table 4-3 Tracking RACI explained

Delivery Toolkit – Tracking RACI									
Project:	*<insert project title>*				**Sponsor:**		*<insert name>*		
Date:	*<insert date>*				**Project Manager:**		*<insert name>*		
Activity tracking									
Activity	**R (responsible)**	**Y/N**	**A (accountable)**	**Y/N**	**C (consulted)**	**Y/N**	**I (informed)**	**Y/N**	**Gap analysis**
<insert activity>	*<insert name>*	*<yes or no>*	*<insert name>*	*<yes or no>*	*<insert name>*	*<yes or no>*	*<insert name>*	*<yes or no>*	*<insert comment on any gaps>*
Action plan									
<insert any actions required as a result of the gap analysis>									

Activity tracking

Each activity and name from the original RACI Chart is inserted, and then a review is conducted and any gaps noted:

- Did the appropriate person (R) complete the activity?
- Was appropriate approval achieved (A), if appropriate?
- Were appropriate team members or external stakeholders consulted (C)?
- Were all appropriate people kept informed (I) of the progress and completion of the activity?

For large projects it may be appropriate to only review activities that are high risk. If an inappropriately qualified or competent person completes the activities, then the project is at risk (due to either lack of regulatory compliance or lack of performance).

Action plan

For each identified gap, an impact assessment should be done, and based on that any required actions should be progressed. For example, if a document required a specifically qualified person to approve it and that has not been done, the action must be to complete the required review and approval. On the other hand, if an activity involved an end-user representative other than that planned, it may be a simple case of agreeing with the customer that the person involved was appropriate and no other action is needed.

Identifying and resolving team issues

All teams will have some issues when working together and a key aspect of project management is to address those issues before they become problems.

There are two tools proposed in Table 4-4 and Table 4-5 which can be used to facilitate this. The first, the Team Issues Tool, is essentially a way for the team members to highlight issues and allows the Project Manager to understand those of the highest priority. The second tool, the Project Team Audit Tool, is intended for the Project Manager, to evaluate the effectiveness of the Project Team and establish resolution of the problems.

Table 4-4 The Team Issues Tool explained

Delivery Toolkit – Team Issues Tool		
Project: <insert project title>	**Sponsor:**	<insert name>
Date: <insert date>	**Project Manager:**	<insert name>
Team issues identification and analysis		
Team issue	**Team score**	**Pareto score**
<insert team issue>	<insert total score from all team members>	<insert Pareto percentage>
Team action plan		
<attach pareto analysis as a histogram as per example Figure 4–7 and insert actions to be progressed to solve the issues with a cumulative pareto score of 80%>		

Tool: Team Issues Tool

The aim of this tool (Table 4-4) is to ensure that the Project Team have an opportunity to discuss and resolve issues which are preventing team performance.

Team issues identification and analysis

This is done in two phases:

1. **Team issue** – the team take part in a traditional brainstorm process:
   - The members are asked to write down any issues they see with current team performance on post-it notes.
   - Once everyone has finished writing the post-its are all put up on the wall.
   - The team work together to categorize the issues by collating similar post-its.
   - The final list of issues is written onto a flip chart.
   The types of issues people may introduce cover all aspects of team working, for example:
   - Conflict or apathy in the team; the team culture.

- Problems with virtual team working.
- Changing team members or the Project Manager.
- How performance is measured and communicated.

2. **Score** – the team are then asked to score each issue on the flip chart (the main categories). This can be done in a number of ways:

- Ranking score – Ask each team member to rank the list in terms of importance of resolution to them personally. The score for each issue is then the sum of the rankings. The lowest scoring issue is therefore the top one, requiring resolution.
- Voting score – Give each team member three votes (decision on the number of votes depends on the total number of team members taking part and the total number of issues identified). The score for each issue is then the sum of the votes. The highest scoring issue is therefore the top one, requiring resolution.

The score can be converted into a percentage score, thus allowing a Pareto analysis to be conducted. The aim of the Pareto chart is to establish the top issues which take up 80% of the total score. An example is shown in Figure 4-7.

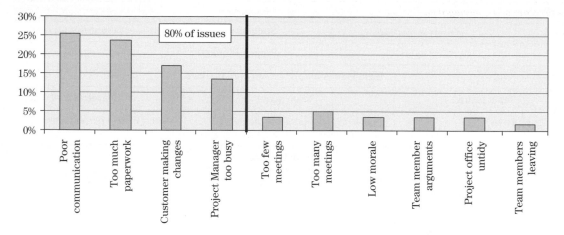

Figure 4-7 Example of a team issue Pareto chart

Team action plan

The aim of the team action plan is to put a plan in place which will solve approximately 80% of the issues, on the basis that there is a diminishing return in solving all issues. In the example in Figure 4-7, 10 issues were highlighted by a team, with the top 4 obtaining 80% of the score. Thus the team focused on these 4 issues to generate a step change in team performance.

Tool: Project Team Audit Tool

The aim of this tool (Table 4-5) is to review the effectiveness of the team – their ways of working (WoW) and general performance. This would usually be conducted informally by the Project Manager or more formally by the sponsor or another senior stakeholder. The key goal for this type of audit is to identify problems early so that project success is not compromised.

Table 4-5 The Project Team Audit Tool Explained

Delivery Toolkit – Project Team Audit Tool			
Project: <insert project title>		**Sponsor:** <insert name>	
Date: <insert date>		**Project Manager:** <insert name>	
Team checklist			
Check		**Response**	**Comment**
1. *Is team management appropriate?*		*<yes or no>*	*<insert response>*
2. *Is team communication appropriate?*		*<yes or no>*	*<insert response>*
3. *Is the team working area (office or virtual environment) supportive of success?*		*<yes or no>*	*<insert response>*
4. *Is team morale supportive of success? (how the team behaves, its values)*		*<yes or no>*	*<insert response>*
5. *Is the team being managed effectively?*		*<yes or no>*	*<insert response>*
6. *Is the team performing (achieving its goals)?*		*<yes or no>*	*<insert response>*
7. *Is there conflict in the team?*		*<yes or no>*	*<insert response>*
8. *Is the team working effectively? (individual performance, clarity of role and WoW)*		*<yes or no>*	*<insert response>*
Team action plan			
<insert any additional actions to be progressed to support team performance>			

Project type

Whilst each project generally follows the overall roadmap described in Chapter 1 (Figure 1-1), the project type affects the way in which particular stages within the roadmap are handled.

People tend to consider projects in terms of their size, complexity and criticality. For example the level of cost, the duration of schedule and the complexity of scope delivery:

- Is the project being developed or delivered in a single location, or many locations globally?
- Does the project involve the creation of new assets or modification of existing assets?
- Is the project cost less than £1 million or greater than £10 million?
- Will the project be completed in more or less than a year?

These four criteria are not mutually exclusive – a small project can be critical or very complex and a big project can be 'simple'.

An additional dimension to project type is provided by looking at the category the project falls into:

- **Asset projects** – projects which deliver a physical asset (e.g., a new plant, product or piece of software).
- **Research projects** – projects which deliver knowledge (e.g., a due diligence review of a potential acquisition or a clinical trial for a new drug).
- **Change projects** – projects which deliver a transformation to current ways of working (e.g., outsourcing IT support or reengineering a sales process).
- **Relationship projects** – projects which set-up or change a business relationship (e.g., a supply contract between a steel panel supplier and a car manufacturer).

To add more complexity to the picture, some larger projects may include a combination of categories – for example, a major construction project may include elements of research (to identify new knowledge required), relationships (establishing procurement agreements) and asset development (to create the final building).

Although the general project methodologies described in this book and in the other books in the series (page *vii*) are robust and can be adapted for different categories, sizes, complexities and criticalities, there are specific questions that need to be asked which can change the way the project is delivered depending on its type.

The key questions that need to be asked are based on:

- What defines the project scope?
- How will we define, measure and control the deliverables (interim and final)?
- How will we assess project success?

Table 4-6 illustrates the impact of these questions across the different categories of projects.

The difficult task for the Project Manager in managing delivery for these different types of projects is in adapting his frame of reference to the scope and the control levers available in a project.

Table 4-6 Project categories and their impact

Project category	Questions		
	What defines the project scope?	**How will we define, measure and control the deliverables?**	**How will we assess project success?**
Asset	Usually detailed requirements with a defined outcome: ➤ A new facility to manufacture a drug product ➤ A new accounting software system	Against specific milestones and 'hard' deliverables: ➤ Delivery of equipment to site ➤ Approval of software design specification or start of module testing	Did we create what we set out to create? Are the requirements all met? Is the final result delivering what was intended? (this may be an extended evaluation over a significant portion of its useful life)
Research	Usually a problem or an idea: ➤ What will the impact of rising sea levels be on UK flatlands? ➤ What would happen if we remove a wash stage from the process?	Against time or the completion of specific goals: ➤ The study will last six months ➤ Defined pilot experiments to establish if the research is on target	Does subsequent experience show the research to be correct (or incorrect)? Can we initiate (or extend) a project to put the research results into practice?
Change	Usually a problem, an opportunity or a solution presented by others: ➤ Sales order processing is too expensive ➤ Outsourcing IT tasks will improve service levels	Against specific milestones, 'hard' and 'soft' deliverables: ➤ New WoW established ➤ Training complete ➤ Survey of affected people shows satisfaction with the change	Is the new business process sustained? Are we seeing the benefits we anticipated?
Relationship	Usually with a set of requirements or an opportunity: ➤ An external management company is offering to take over all our facilities management ➤ We need to establish a secure source of supply for a raw material to meet our specification	Against specific deliverables: ➤ Complete bid enquiry pack ➤ Acceptable vendor responses	Is the supplier meeting expectations?

Asset projects

These tend to be the easiest to understand in control terms since they have a reasonably well-defined approach (e.g., erecting walls is not possible if the foundations are not complete). Control within this type of project is essentially linked to clarity of funding requirements and about checking hard deliverables and milestone achievements. The roadmap in Figure 4-8 illustrates the main stages and stage gates in this type of project.

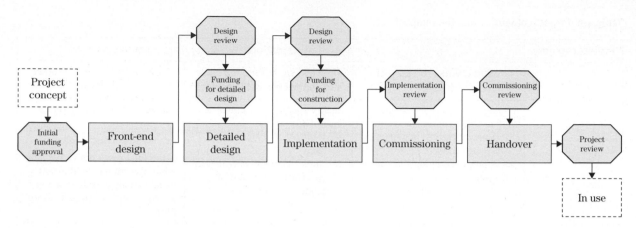

Figure 4-8 Project roadmap – asset project

Research projects

These are challenging because the scope tends to be fuzzy (an idea) and progress can be heavily dependent on individuals. Classical techniques of measuring completion (e.g., how would you know if you are 90% through the formulation of a theory?) are less relevant. This is why a lot of successful research projects tend to run on a time basis with a lot of stage gates (pharmaceutical drug discovery and development may have as many as 8 or 10). Control becomes a matter of establishing clear criteria for the stage gate reviews and an understanding with the Project Team that each of these reviews is a go or no-go decision point. An added complexity is that a research project which returns a negative result is not necessarily a failed project. That negative result has increased knowledge and may suggest new research opportunities therefore the assessment criteria need to consider the potential for negative results. The roadmap in Figure 4-9 illustrates this type of project.

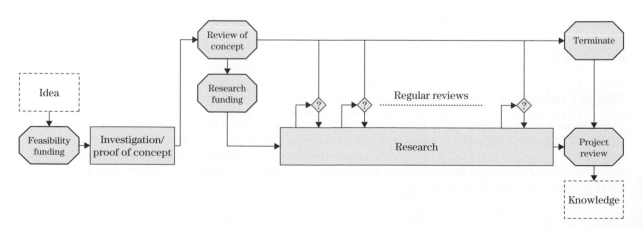

Figure 4-9 Project roadmap – research project

Change projects

These are very much about the impact on people. Whilst certain parts of the project will require consideration of hard deliverables and associated costs, a major part of the control of delivery is stakeholder management and the sustainability of change. No-one will thank you for bringing in a new sales order processing system on time and on cost without ensuring the change to people and their ways of working is sustainable. The checkpoints on this type of project roadmap are shown in Figure 4-10. Note that in this case, the sustainability check is a formal part of the project delivery process, to confirm that the change is robust, and feeds into the project review.

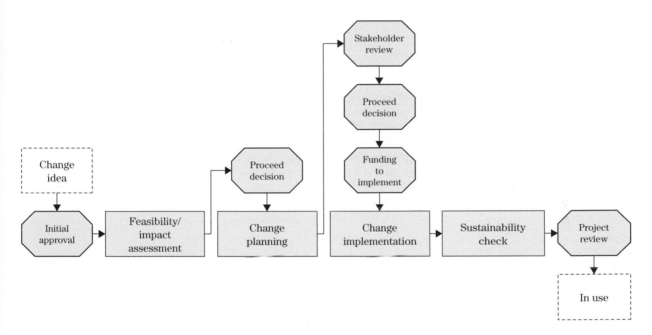

Figure 4-10 Project roadmap – change project

Relationship projects

This is another area where there is a blend of hard deliverables (e.g., a contract) and the softer aspects of stakeholder management. Delivery of this type of project requires not only a focus on the requirements, but potentially considerable stakeholder management to ensure support for the relationship that is being built (e.g., executive level sponsorship between supplier and customer organizations). The roadmap for a relationship project is shown in Figure 4-11.

Figure 4-11 Project roadmap – relationship project

Project roadmap and risk mitigation

Managing different types of project is essentially about managing risk. The number of decision points (or stage gates) within the project and the focus of the Project Manager reflect where the risk is likely to be at its highest. So, in a research project, the risk is that the research continues for extended periods without generating any useful benefits. In a relationship, the risk is that the stakeholders do not agree with the project decision.

Funding strategy and finance management

Organizations will inevitably have their own processes and procedures regarding how projects are funded and how those funds are managed. Adherence to these company 'rules' usually remains the responsibility of the Project Manager, although most organizations would provide appropriate support from their Finance Departments (depending on the level and complexity of the funding). In any event, the Project Manager must maintain an effective link between the project funds and organizational finances.

Project funding

The initial decision to proceed with any project will generally have considered the benefits and costs in isolation from the overall business finances. Project funding is where the company and project finances meet. Most companies have four sources of funding for projects; retained earnings (previous profits), issuing new shares, long-term or short-term borrowing. The ability of a company to undertake projects is therefore limited by its profitability and its creditworthiness. Project funding will have a direct impact on company finances through two basic needs: needing money and needing profit.

Needing money

Projects tend to run on money. If a project is financed by bank loans, then there will be a financing cost that the Project Manager may not see directly. Projects that are delayed or those in which stage payments are not received will see increased financing costs. Disputed or very late payments can have the effect of driving a company out of business. For example, in 1991, Davy McKee became insolvent as a result of delays in a project for a platform for the Emerald oil field, which in turn delayed the payment of $100 million at a time when the company could not refinance the project.

Whilst a Project Manager may be unable to influence the financing strategy, understanding where the money comes from and the commercial impact to the business of changes to payment or schedule is important. In this regard a Project Manager must link cost management with finance management.

Needing profit

This usually affects suppliers or contractors more than end-user companies and can have a significant effect on the business. In the supply of equipment or services, revenue and profit are only realized at the time of delivery or acceptance (the two may not be at the same point). What that means in practice is that if a company supplying a piece of equipment cannot deliver due to a delay in their customer's project, then they cannot claim profit on the 'sale' even if they have been paid in full. What tends to happen in this case is that the customer will be asked to take ownership (or title) for the equipment and, most importantly, the associated risk (if the factory burns down, it is the receiver's loss).

Another area where profit can be impacted by a change in project schedule is when it pushes delivery over a financial year end. This might be acceptable for the project, but the company may not find it acceptable. In simple terms, it would mean that the revenue and profit forecast in one year moves to the next. The internal pressure that this generates on a Project Manager should not be underestimated. The failure of a company to deliver on its financial promises can be difficult for all and will inevitably result in significant project management effort being expended on trying to recover the situation. On rare occasions, the company may need to slip revenue and profit out of a good period into the next. Again the project management effort in manipulating the project to achieve this can be significant.

Project and company finances

Business accounting is very different to project accounts (or cost control). Project accounts are there to support the decisions the Project Manager needs to make about the financial position of the project. Business accounts provide a mechanism to support the overall management of the business, and there are specific rules governing the way in which these accounts are maintained and reported; many of them having statutory authority.

There are two types of business account, financial and management:

1. *Financial accounts* are generally those which appear in company reports and must be made publicly available by law. They show the financial standing of the company at the time they were prepared, and since they are published, they show the minimum information to satisfy the law.
2. *Management accounts* are usually prepared on a monthly basis and are far more detailed, in that they will break down the business into its lowest reasonable level. Frequently management accounts are produced on a business unit basis, and it is here that the project may have some visibility.

Management and financial accounts have the same structure and comprise a balance sheet and a profit and loss (P&L) account. The balance sheet records the assets and liabilities of a company. Whilst this is a useful view of the company's strength and stability, the P&L account is regarded as a more instant view, provided the reader understands that:

- The P&L account is a snapshot in time. It shows the financial performance of the company at a particular point. It is not a forecasting tool (unlike project cost control reports).
- Profit or loss is not the same as cash, and the time at which items are entered on the P&L account do not necessarily coincide with the payment and receipt of cash. For example, sales are usually entered at the point of delivery and invoice. The payment period for the invoice could be up to 120 days. The outstanding invoice becomes a trade debt, but profit can be declared well before the money is received. The obvious result of this approach is that a company can declare significant profits and have no cash.
- The cost of capital assets does not appear in the P&L account as the value spent. What appears on the P&L is the cost associated with any interest payment on borrowed money (for the expenditure) and the depreciation charge of the capital (the amount by which the asset reduces in value over time).
- Funds for revenue projects (projects which do not generate a capital asset) show directly in the P&L (as revenue means money generated by the business) when the money is spent.

Matching project and company finances

Within a project, there are generally three types of money that the Project Manager looks after:

1. *Budget and contingency* – what the project should cost.
2. *Committed* – what you have committed to spend (usually orders placed or man-hours consumed).
3. *Forecast* – what the final cost of the project will be.

Business finances will directly 'see' the following:

➤ *Expenditure* – money that has actually left the company.
➤ *Committed* – project commitments in terms of orders placed (or hours spent).

It will not see the budget or forecast unless the Project Manager provides the linkage. This is usually done through two approaches:

➤ **Cash-flow forecasts** – a forecast of when money will leave the business. This can be particularly important when the project crosses over end of year accounting periods.
➤ **Accruals** – a forecast of business commitments which will result in a payment that goes over the financial year end.

Figure 4-12 illustrates the relationship between the various components that make up the financial picture for both the project and the business. Whilst a Project Manager may manage the project accounts through tracking how funds are committed, the finance department are likely to track only expenditure. In reality, the Project Manager would keep an overview of both situations through a cumulative cost curve.

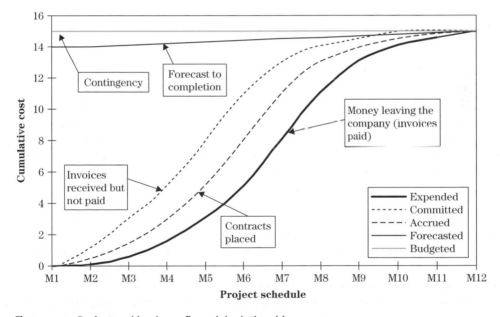

Figure 4-12 Project and business financial relationships

Explaining the cumulative cost curve

The five cumulative cost curves in Figure 4-12 are further explained in the following:

➤ **Budgeted** – the approved project funds; typically this level of funding cannot be exceeded without further approval from the organization.
➤ **Forecasted** – the projected total project spend. Note that the difference between the forecast and the total budget is usually the contingency, which is not forecast for use.

➡ **Committed** – the project funds that have been committed through placement of contracts/ purchase orders for goods or services. The project funds are effectively set aside for these specific purchases, even though the money will not leave the organization until the goods or services are received.

➡ **Accrued** – the anticipated payment of invoices received against specific purchase orders. Usually once an invoice is received and approved, the company considers the funding as accrued until it is expended according to agreed payment terms.

➡ **Expended** – the payment of approved invoices against specific purchase orders. Money has effectively left the company at this point.

Cash flow is very important in ensuring that funds are made available for the project at the right time. Accruals are important in that they allow for money committed in one financial year to be accounted for in that year. On revenue funded projects, they are critical as money not spent in a particular year is effectively 'lost' and may result in an overspend in a subsequent year.

The business will therefore be very interested in the expended and accrued curves in Figure 4-12 and the Project Manager needs to ensure that the speed of commitment of project funding will deliver the appropriate expenditure profile to match organizational needs. A common issue is when a project has to either speed up or slow down expenditure to meet these needs. This can be achieved through management of order placement or invoice management.

Placement of purchase orders

Either delaying or speeding up placement can have an adverse impact on a project.

➡ *Delay* – The supplier may not be under any obligation to retain the agreed price or delivery terms. This is often overcome through the agreement of quotations for a 3 to 6 month validity.

➡ *Early* – Often, purchase orders or contracts placed earlier than anticipated are not backed up with the appropriate level of design or specification. Therefore to be sure that an incorrect order is not placed, the whole of the work associated with that order has to be fast-tracked (taken out of sequence).

Management of invoices

➡ *Invoice issue delay* – Through agreement with the supplier the invoice could be issued later than agreed within the contract. This is a risk to the supplier as some work has already been completed for which payment cannot be demanded.

➡ *Invoice issue early* – Through agreement with the supplier, the invoice could be issued earlier than agreed within the contract. This is a risk for the organization as payment request is accepted for goods or services not actually received.

➡ *Invoice paid late* – Ideally through agreement with the supplier but often not. The organization may choose to pay an invoice late to suit their financial management needs.

➡ *Invoice paid early* – The organization may choose to pay an invoice early to suit their financial management needs. This is a significant risk for an organization, particularly when it is in combination with an early invoice for goods or services not yet received.

In any of the above situations the impact on the supplier, the organization and the project needs to be assessed.

Getting paid and paying for things

Paying for things and getting paid is another area where the project and business meet. The following topics would normally all be covered within the contract or purchase order for goods or services being procured and can often form a crucial part of the negotiation process.

Payment terms

The first and most important thing is to know what procedures the accounts department uses. Deviating from these will be difficult and require further time and effort to ensure that things happen when the project wants them to.

Typical payment terms that appear on orders are 'nett monthly,' '30 days' or '60 days'. *Nett monthly* means the invoice will be paid in the month after the receipt of the invoice, not 1 month after the invoice. Thirty days and 60 days mean what they say, the invoice will be paid 30 or 60 days after receipt. Many companies operate on the nett monthly approach and have a set day for invoice payment. If that day is missed, then the payment will slip for a complete month. Do not, therefore, promise payment terms that cannot be delivered, and ensure that any purchase enquiries and orders include details of the organization's payment terms.

Payment milestones

It is important that the milestones for payment are clear, unambiguous and not open to interpretation. So, for example, payment on ex-works delivery should mean just that, not payment on arrival on site or after certain documentation has been received. A Project Manager will have an influence on issuing an invoice at the appropriate time and for approving payment for purchased goods. There is nothing more stressful than attracting the attention of the credit controller because an invoice is overdue or the attention of a supplier's credit controller because an invoice has not been paid.

Retentions

Retentions are where a proportion of the contract, usually 5–10%, is invoiced, but the payment only becomes outstanding at a later date, usually the end of a warranty period.

Guarantees

A bank guarantee is where money is received (usually as part of a performance or advance payment), but the supplier provides a guarantee for the payment against a specific time (to delivery of the goods) or time period (end of the warranty).

If guarantees are to be used, then the finance department must be consulted as most organizations have a standard form of words for a bank guarantee. In many cases, the bank will need to approve the format before issuing. The major points to be aware of are:

- The guarantee is available for a specific reason and is 'on demand'. This means that the guarantee can be called by the holder without notification.
- The guarantee is limited in value and duration. A statement needs to be included on the expiry of the guarantee because they tend to be things that, go missing, after the project is complete.

Banks do charge for setting up a guarantee and then charge a percentage, typically 1–2% per annum. Guarantees have a direct impact on company finances as they are a financial liability. So whilst they seem attractive to help with cash flow in a project, they may not be desirable for the company.

Doing business overseas

It may be possible to source suppliers, nationally but sometimes it will be necessary to procure from, ship to, install and pay for equipment overseas. In conducting business overseas, there are some key areas that need to be considered:

Currency exchange

Operating a project in foreign currencies introduces complexity and can have a significant impact on its viability. There are a number of approaches, some of which depend on the size of the project:

➤ A Project Manager should use only one (preferably his organization's national) currency. If this is not possible, nominate the project currency, preferably one that is stable and international. Request that suppliers quote and deal in the nominated currency (they take the risk) but be aware that they will have included a price for this risk or limited the variations in currency. Evaluate the risk and benefit of this approach as for any other project cost.
➤ Buy currency ahead (hedge). This is likely to be relevant for large projects and an area where the finance (or treasury) department of the company will be involved. The advantage of this approach is that it fixes the exchange rate for the project life.
➤ If a foreign currency must be used, a Project Manager should be familiar with his company's policies. He must not agree something with a supplier or client that the organization does not allow.

Where a project must deal with more than one currency, the best approach is to set a project 'rate' and convert all costs to local currency at this rate. This will prevent currency fluctuations distorting true project costs in cost control reports. Currency fluctuations can then be accounted for as 'below the line' costs so it is clear where change is a consequence of currency fluctuation and where it is a consequence of true project change.

Movement of goods

Once goods have been purchased, it may be necessary to move them across national borders. For some parts of the world, harmonization of customs rules means that the movement of goods is relatively simple. For other parts of the world, there is almost always duty to pay and potentially comprehensive and specialized paperwork. These will normally include a packing list and shipping invoice as well as a certificate of origin, which needs to be attested by an official body such as the Chamber of Commerce. Key points to bear in mind are:

➤ Has the duty been considered in the costing? (5% is typical, although for some countries and equipment 12% or higher is common.)
➤ The shipping paperwork can only be done when the goods are packed and ready for shipment and may take 1–2 weeks. Has this time been allowed in the project plan?
➤ What are the customs requirements and who will handle them?

➤ What are the restrictions on shipping material? In some cases it is impossible to ship certain types of goods (technology or certain chemicals) into certain countries. In other cases, some materials have restrictions on their shipment method. For example, bulk chemicals, touch-up paint or lead-acid batteries (perhaps in a back-up power supply) cannot be shipped by air and untreated wooden packing crates cannot be shipped into some countries.

Professional assistance from either within the organization or a specialist shipping company is almost always worthwhile.

Getting paid

Many of the criteria for payment and getting paid overseas are similar to those previously discussed, though there may be a need for specialised methods of payment. One such method is a documentary letter of credit.

A 'letter of credit' is a bank instrument which guarantees that an amount of money is lodged with a mutually agreed bank and that this money will be paid on demand with suitable shipping paperwork. The shipping paperwork is normally the shipping invoice and the bill of lading for the ship.

There are two forms of letter of credit: revocable and irrevocable. A revocable letter of credit can be cancelled or amended at any time without the exporter being notified (and is therefore of limited use). An irrevocable letter of credit can be amended or withdrawn only if all parties agree. An irrevocable letter of credit can also be 'confirmed' or 'unconfirmed'. If it is 'confirmed', then the bank has added its own undertaking to pay under the terms of the letter. An 'unconfirmed' letter means that the bank will not undertake to pay and merely notes the letter.

To safely conduct business, an irrevocable, confirmed letter of credit is required. Without this, the letter of credit may not be honoured.

Liaison between project and accounts

Good liaison between project financial controls and the business is critical in order to successfully complete all but the simplest projects. The project accountant (whether full time or part-time) is an important part of that link. A good working relationship between Project Manager and project accountant will ensure that the rules are not inadvertently broken the importance of business financial controls forgotten.

Asset management

Apart from the management of cash flow, year end accruals and overall expenditure, the finance department will be interested in how the project is impacting company assets. Most project types can dispose of assets, amend them or generate new ones, and each needs to be tracked within the organizational finances. Assets are the way that the company generates revenue, and so any change needs to be tracked within the company balance sheet. Typically the timing of asset changes is a key area of concern for a company:

➤ *Asset disposal* – When an asset is disposed of, it is no longer generating any revenue for a company and will therefore not depreciate any further.

➤ *Asset generation* – New assets need to be maintained, and this represents a new cost to the business which is seen in the company finances rather than the project finances. Project finances only consider the initial cost of acquiring the asset so that it can start generating revenue for the company.

Financial governance

Often the most crucial role of the finance department is the provision of appropriate advice to the project relating to financial rules and regulations under which the company must operate. Whilst large projects may have a dedicated project accountant who liaises with the company's financ management team, it is often the case that the Project Manager must provide that strong link to ensure financial governance.

Scope management and delivery

Project scope defines the way in which the project will deliver benefit to the business. Scope can evolve over the life of a project, and one of the Project Manager's greatest challenges is the need to ensure that the project scope continues to deliver value.

Scope management and delivery should not be confused with change control – the first is about managing the delivery of value, whilst the latter is about managing cost, quality and time. Formal control of change is covered in Chapter 5, whilst this section covers scope management and delivery. Being in control of value delivery is at the core of being in control of the project. Scope should be viewed in terms of its ability to enable benefits and so scope management is, to a large extent, the management of benefits realization (Figure 4-13):

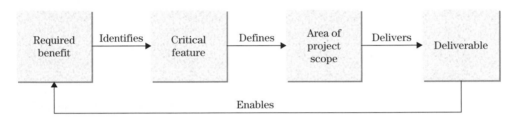

Figure 4-13 Linking scope to benefits

Scope management process

Remember that the Project Manager's role is to manage (scope) quantity, quality and functionality to deliver value. The process of scope management is really about value control and value delivery (Figure 4-14).

Value control

The scope of a project is developed and defined so that each activity or deliverable required for success is included. Scope which does not contribute to the outcome is eliminated, and the remaining scope has a clear link to the benefits to be realized from the project (Figure 4-13). The goal then, of value control, is to ensure that this scope is delivered accurately and effectively. The process for value control is shown on the left hand side of Figure 4-14:

Identify scope change

During delivery, there are times when scope changes are identified. These instances are discussed in more detail later in this section. Whatever the reason for the proposed change, the key is to link it back to the scope definition. A variety of tools are available including the following:

From *Real Project Planning* (Melton, 2008):

➤ *Path of Critical Success Factors (CSFs)* – In order to achieve project success, a series of CSFs are identified and defined. A CSF is an identifiable action/activity that is required to achieve project success. This can be tabulated in a table of CSFs.

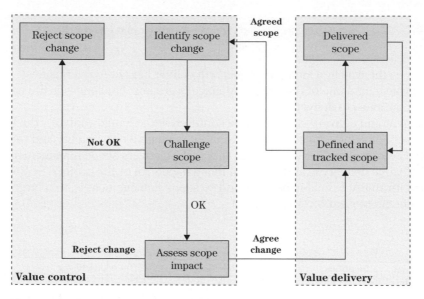

Figure 4-14 Scope management process

- *Critical to quality (CTQ) tree* – A CTQ tree is a diagrammatic representation of the link between critical features and scope. A critical feature is a feature which the business uses to assess project success and is generated through the consideration of the benefits which are required by the business.
- *Work breakdown structure* – A way to segment the scope into bounded activities and deliverables and usually combined with the definition of CSFs.
- *Quality matrix* – A way to define interim deliverables so that quality, quantity and functionality are defined.
- *Activity plan* – A way to formally align project activities with CSFs defining quality, quantity and functionality as well as internal dependencies between activities.

From *Project Benefits Management* (Melton et al., 2008):

- *Critical to Quality CTQ Scope Definition Tool* – This is a tool which tabulates a CTQ tree and in effect provides simultaneous scope definition and challenge. If there is no link from a benefit to a critical feature to an area of scope to a deliverable (Figure 4-13), then the deliverable and/or scope is not a part of the project value proposition.
- *Hierarchy of objectives* – a diagrammatic way to link a benefits map to a project scope. It allows the linkage of project objectives to business objectives.

Each can be used as the basis for assessing scope delivery progress and therefore the basis for scope change assessment.

Challenge scope

Each proposed change of scope should be subjected to an objective challenge which links back to the value proposition for the project. In other words, 'Without the scope change, can all critical features be delivered?' No scope should be added, deleted or modified such that the benefits case is not capable of being delivered in full. The Scope Tracker (Table 4-7) addresses this challenge.

Assess scope impact

Each value-add scope change needs to be delivered within a project which is already in progress, therefore the impact of the change needs to be assessed and the risk to the project outcome identified. It may be that the scope change is required in order to deliver the complete value but its delivery would cause too much risk to other elements of the value delivery. In other words, 'With the scope change, is the delivery of any critical feature adversely impacted?' Again, the Scope Tracker (Table 4-7) addresses this challenge.

Reject scope change

Scope changes which have no value or are too disruptive to implement should be rejected. Communication of these rejections needs to be clear as scope changes can typically involve a lot of stakeholders with varying levels of influence and power.

Value delivery

During project delivery value is realized continuously as the scope quantity, quality and functionality is delivered through completion of activities, interim and final deliverables. The process for value delivery is shown in the right hand side of Figure 4-14:

Defined and tracked scope

All the scope definition tools listed on page 111 can be used to track the progress of scope delivery. For example:

➤ CSF delivery can be tracked against milestones within the project roadmap.
➤ Deliverables can be tracked against the stages of development as defined within the quality matrix.

Delivered scope

As scope is delivered and final deliverables are completed, benefits realization should be enabled. This is done through the delivery of critical features of the project.

Scope change

Changes to project scope happen via two routes: externally through customer, supplier or environmental effects and internally through the Project Team. Scope changes come in three general types:

➤ Requirements change.
➤ Cost change.
➤ Scope creep.

Requirements change

In certain circumstances, a change in requirements is necessary for project success. For example, a manufacturing plant may need an additional 10% capacity to cope with emerging market forecast sales for a new product.

A change in requirements is generally one of the easier scope changes that the Project Manager deals with (assuming that the requirements are adequately defined in the first place). They are generally clear and more easily allow of a formal assessment of the impact on the deliverables. They are often presented in terms of the business case by senior stakeholders and therefore frequently have access to the additional resources needed (funds, people, assets).

Cost change

Too often, a change in the cost of a project is necessary for project success. For example, a change in the external market may change the business case leading to reduced funds available for the project. Scope changes driven by cost considerations can have a major impact on the project unless the change is objectively challenged against the critical features of the project. In this scenario, the Project Manager must ensure that the stakeholders buy-in to the change in value delivery (as impacted by the reduction in funding). Too often project funding is cut but expectation of project output remains the same.

Scope creep

Scope creep is an extremely common cause of scope change. This is the incremental change of scope over time, typically occurring in an ad hoc manner. For example Project Team members, customers and other stakeholders suggest or imply change in a low-key manner: 'Why don't we also…?'

Project scope creep can threaten the project's ability to deliver and comes in three main types:

- **Requirements creep** – usually where the initial requirements are added to in a piecemeal way. For example, a customer may 'fill in the blanks' with additional requirements that were not clear in the original scope. The only way to mitigate this situation is to ensure that the scope is fully defined in the first place and that delivery does not commence until the scope has been challenged against the value proposition.
- **Feature creep** – often associated with software projects, but also occurs in many technology lead activities. Where the Project Team adds functions which were not explicitly asked for, but are seen as 'useful', 'will be needed' or 'just in case'. The most appropriate way to mitigate this situation is to challenge each additional function against the critical features of the project.
- **Instruction creep** – associated with change or procedural projects, where, for example, the initial simplification of business instructions is made more complex to cater for unforeseen eventualities (analogous to the idea of a camel being a horse designed by a committee). Management and minimization of this type of scope creep is entirely due to good stakeholder management on the part of the Project Manager, sponsor and the Project Team.

Short case study – feature creep

In the mid-1990s, Apple Computers was looking to replace the operating system for its Mac computers with an updated version under a project codenamed *Copland*. This was to be a radical update of the existing System 7.5, with backward compatibility and a host of new features – including native support for a new chip set. As this project gathered momentum within Apple and became a priority, existing projects and development teams began to add to the scope of Copland to ensure that they maintained their existence. In addition, when external developers saw early versions of Copland, there was pressure to add yet more functionality. The project became ever more complex, and the new operating system's delivery date continued to slip. By 1997, it was obvious that the project in its current form could not deliver a stable product, and it was cancelled. Apple subsequently acquired NeXT and used that company's technology to create the Mac OS X operating system. In this example, the project paid the ultimate price for not managing the feature creep; it was stopped and substituted.

Tool: Scope Tracker

The CTQ Scope Definition Tool introduced in *Project Benefits Management* (Melton et al., 2008) is the basis for the Scope Tracker (Table 4-7), which is used to identify, challenge, assess and track scope changes. Changes to project scope are most easily identifiable through changes in deliverables, and this tool links the deliverables to the project scope and then to the critical features and finally the benefits.

Table 4-7 The Scope Tracker explained

Delivery Toolkit – Scope Tracker							
Project: <insert project title>				**Project Manager:** <insert name>			
Date: <insert date>				**Page:** <1 of 1>			
Deliverable	**Change?**	**Scope**	**Impact?**	**Critical feature**	**Impact**	**Benefit criteria**	**Impact**
<insert agreed output from the project>	<insert change – add, delete, modify>	<insert agreed scope to deliver the CTQ feature>	<insert impact of change to deliverable on scope>	<insert a CTQ feature>	<insert impact of scope change on CTQ feature>	<insert organizational or project benefit>	<insert impact of CTQ feature change on benefit>
Scope change summary							
<insert summary status of scope changes – accept, reject, refer for further development>							

Deliverable

Any scope change whether by requirement change, cost or scope will have an impact on project deliverables. Deliverables which are to change due to the scope change should be listed here.

Change

The agreed deliverables for the project should be assessed and the nature of the change identified; either an additional deliverable or the modification/removal of an existing one.

Scope

The original scope area which the deliverable was linked to should be identified here.

Impact (on scope)

A changed deliverable will have an impact on the scope, and this is where the first assessment of the scope impact will be identified. If the deliverable changes, what is the impact on the scope area?

Critical feature

The original critical feature which the scope is linked to should be identified here.

Impact (on critical feature)

The change in scope will have an impact on a critical feature, and at this point, the assessment process will be able to identify what the change is. Scope creep, in particular, can be identified at this point, usually through the fact that the scope change has no impact on existing critical features (but may add new ones).

Benefit criteria

The original benefit feature which the critical feature is linked to should be identified here.

Impact (on benefit criteria)

The final part of the tracker assesses the change to critical features against the agreed project benefits. Usually the critical features will be linked to general categories of benefits rather than specific benefits metrics. Often, it requires a number of project deliverables to enable a set of specified benefits in any one category or criteria.

Short case study

During an operational improvement project, two key deliverables are assessed for change during delivery in order to check progress of the scope and hence value delivery (Table 4-8). The business had indicated that previously approved funds were unlikely to be released due to a reprioritization of business needs. The project scope was defined as the change in a specific production area to reduce operating costs and relied on two CSFs:

- CSF 1 – changed ways of working. For example, changing the way that the operators interacted with the production process; how they moved around the plan so that people were used more effectively.
- CSF 2 – changed production process. For example, changing the production equipment itself so that the process was more effective.

The Scope Tracker (Table 4-8) identified that the business case had changed (less cost available so less benefit to be realized) and that if the change in scope was to be accepted, then the sponsor had to accept the change in the business case and champion that with other senior stakeholders. In addition, the path of CSFs had altered along with the overall value proposition.

Table 4-8 Example Scope Tracker

Delivery Toolkit – Scope Tracker							
Project:	Operation improvement			**Project Manager:**		Peter Piper	
Date:	Week 5			**Page:**		1 of 1	
Deliverable	**Change?**	**Scope**	**Impact?**	**Critical feature**	**Impact**	**Benefit criteria**	**Impact**
Cell-based operation layout New standard operating procedure	None	Improve the process cycle time	None	Reduce number of operating personnel	None	Reduce operating costs per unit output	Reduction will no longer meet target, benefits case needs to be reassessed. The benefits are only possible if the equipment and changed WoW are BOTH delivered.
New equipment item A	Cancel on cost grounds	Improve the robustness of the process	No change in robustness possible	Decrease waste from the operation	Waste will be unchanged		
Scope change summary							
Change to scope to reduce project costs accepted, but business case now needs to be reevaluated to confirm that the project is viable							

Value-added scope

In the example in Table 4-8, the impact and the challenge process is relatively simple. For more subtle changes, a formal process around the concept of value and 'lean' scope is required. One of the primary goals of any lean deliverable is to put into the project only what is needed to meet the requirements. Anything else is regarded as waste. Once the scope change has been identified, it can then be challenged using seven wastes (Table 4-9).

The seven wastes are a description of the seven areas where waste can be generated in any type of process. However, Table 4-9 specifically identifies waste definition within the scope management process – specifically scope change. Each time an area of scope is changed, it should be challenged against the seven wastes in order to ensure that value is being appropriately managed.

Once the scope has been challenged, if accepted, the appropriate parts of the scope tracker should be updated to show the new scope. At this point, it may be necessary to work with the project stakeholders and others to ensure that the scope change is considered acceptable in the wider project sense.

Table 4-9 Scope challenge using seven wastes

Waste area	Scope change impact	Example
Waiting	Change can delay decisions or increase dependencies between scope items	A new control system could have an additional interface added for linking to a business system. The new interface will delay the completion of the control system part of the project and create a link to the business system update project.
Transport	Change can require deliverables to be moved excessive distances to be worked on	A manufacturer for a skid-mounted piece of equipment could be given responsibility for integration with another package. This will require the skid equipment to be moved to a staging location in addition to the customer site.
Inventory	Change can increase the number of deliverables, but not change the critical features or benefits	A software project could add a reporting feature to allow the user to generate ad-hoc reports in the future
Overprocessing	Change can increase the need for handling, reviewing and approving deliverables	A pharmaceutical manufacturer requires additional commissioning documentation to be produced to support their internal processes beyond that agreed in the project plan
Motion	Change can require Project Team members to relocate/move within different locations	A change to sourcing technology for a project requires technical experts to travel to India frequently
Defects	Change can increase the opportunity for errors or require rework	A ventilation system package for a plant could be changed from off-the-shelf to bespoke design and build, thus increasing the risk of errors
Overproduction	Change can offer no benefit to the project	The change in a business process may deliver no additional benefits

Set-up plan delivery case study – purchasing business process improvement

To illustrate the key points from this chapter, the delivery phase of a business process improvement project follows. It focuses on the delivery of the set-up plan as outlined in *Real Project Planning* (Melton, 2008).

Situation

A recent review of an organization's ability to purchase key services or equipment has been completed, and the metrics demonstrate that significant improvement is necessary.

- It can take extended periods to select a supplier and then even longer to get a purchasing contract agreed.
- The purchasing team has grown, but this is making little impact on the backlog of purchase contracts yet to be completed.
- The number of contracts that have expired is increasing.

Initial project approval

The business case development proved that unless the business process was improved, the organization would be unlikely to demonstrate appropriate controls of its external spend (a budget of approximately $100 million/year). A Project Manager was assigned and asked to deliver the business case as summarized by the benefits hierarchy (Figure 4-15).

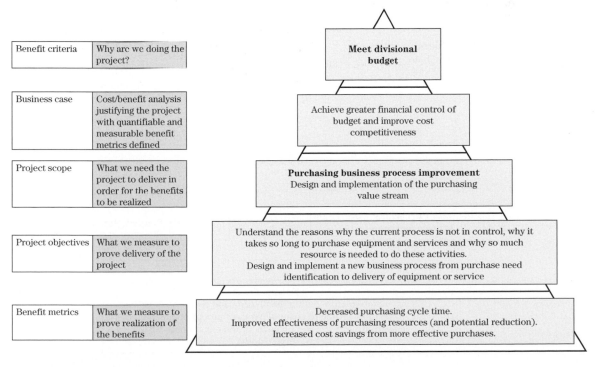

Figure 4-15 Benefits hierarchy – purchasing business process improvement

The benefits required the delivery of a new business process which would add additional value to the business in three distinct ways:

⫸ Get better at doing 'deals'.
⫸ Do the 'deals' faster.
⫸ Spend less effort in doing 'deals'.

It was, therefore, seen as a challenging project requiring a new approach within a fast-track timescale: benefits needed to be delivered within 6 months to support the achievement of the end of year budget targets and other non-financial objectives.

Project set-up plan development

The Project Manager had not completed a project within a purchasing environment previously; however, she felt that her background in generic business change projects was sufficient to cope with the challenge. In addition, she was trained in the use of lean Six Sigma (LSS) techniques, and based on the data presented in the business case, she felt that the use of such techniques would be crucial in quickly identifying and implementing the appropriate business process changes. Following assignment to the project, she quickly set about forming a team and started to plan the delivery of the project. For this reason, she used the development of a set-up plan as a structure for effective project launch.

The set-up plan outlined the project organization (Figure 4-16) and followed the traditional roadmap for a business change project (Figure 4-17).

Both were designed to minimize the use of resources so that operation of business as usual (BAU) was minimally impacted by the early stages of the project. Team members were selected according to a set of criteria which aimed to maximize sustainable success.

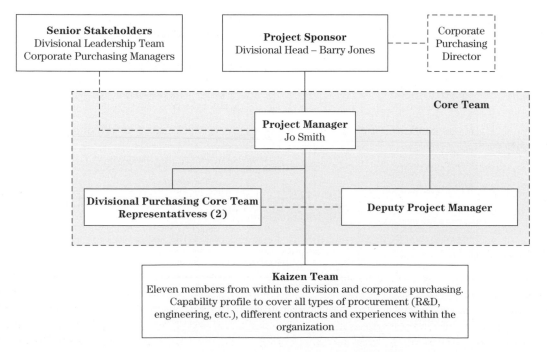

Figure 4-16 Project organization – purchasing business process improvement

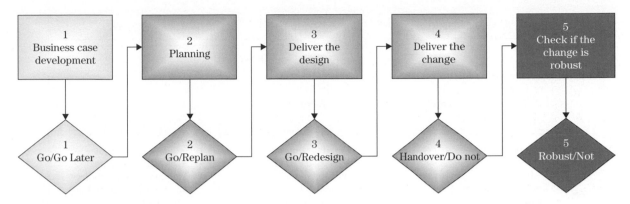

Figure 4-17 Project roadmap – purchasing business process improvement

Set-up plan delivery

The delivery of the project through stages 2, 3 and 4 (Figure 4-17) generally proceeded according to plan, although key issues were seen:

- *Project roadmap* – The delivery approach proposed was novel to all team members except the Project Manager.
- *Project organization* – The pressures of maintaining BAU and delivering the project were seen as issues by team members (regarding having sufficient capability and capacity to deliver such a novel project).
- *Scope management* – The management of a variety of stakeholders required a clear and communicated focus on the CTQ scope and associated critical features.

No general issues were seen with the management of finance as the project was funded through the divisional revenue budget and only required the Project Manager to follow usual approval processes for this funding:

- Develop a resource plan which highlights any loss to the business through not having the resource available for BAU.
- Develop an expense plan linked to the logistics of running a week-long workshop.

Once approved (through a system of allocating approved funds to different cost codes) the Project Manager had authority to commit funding in line with the plan.

Project roadmap

Upon approval of stage gate 2 (Figure 4-17), the Project Manager commenced the design stage. In terms of the kaizen approach (Figure 4-18), this meant completion of the design of the kaizen event and gaining approval from all stakeholders that the event should go ahead. A kaizen event is a resource-intensive but short-time-scale methodology which requires stakeholder and sponsor involvement at the end of each day in a week-long programme (formal decision-making) – in effect, there are stage gates at the end of each day which collate to form the final approval of the output.

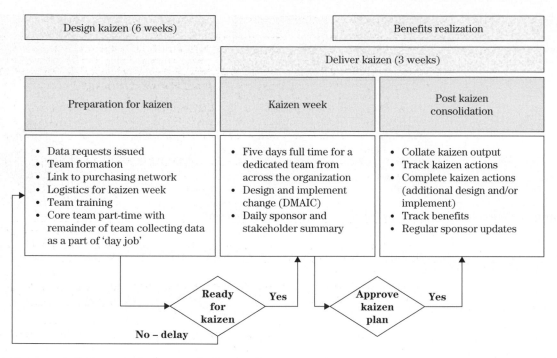

Figure 4-18 Delivery roadmap – purchasing business process improvement

As the majority of the project cost and disruption to BAU occurred during the kaizen week, this approval pre-kaizen was seen as the most crucial. A pre-kaizen checklist was generated to ensure that this stage gate would be achieved (Table 4-10). The checklist focused on the people, scope and set-up risks as identified in the Path of Success. As a result, the kaizen went ahead as originally scheduled.

Project organization

As the kaizen checklist was being completed during week 6 of the project, a series of additional team-effectiveness reviews were conducted:

- Project team effectiveness via review of critical parts of the Project RACI (Table 4-11). This demonstrated that the team were operating as intended; using the purchasing business expertise where applicable and using the business improvement expertise where applicable.
- Project team effectiveness via a team audit (Table 4-12) conducted by the Project *Manager*. This demonstrated that the team were finding it difficult to be involved part-time in the lead up to the kaizen and were sceptical of this event in terms of what could be achieved in only a week.

The tracking RACI demonstrated that data and team capability will continue to be the major challenges during the kaizen week. In addition, the feedback from the team and the senior stakeholders demonstrated that expectations had been raised substantially even if people in the business remained sceptical. Managing such raised expectations became a priority for the Project Manager as she prepared for and delivered the kaizen week.

Table 4-10 Pre-kaizen checklist – purchasing business process improvement

	CSF	Target – Pre-kaizen	Status – Pre-kaizen
1	Robust data	⇒ Ability to demonstrate the baseline performance in terms of the operation of the purchasing business process	AMBER Not all data is available but sufficient to baseline current performance
2	Capable Project Team	⇒ 90% involved in current purchasing business process ⇒ All committed to being at the kaizen week 100% of the time ⇒ All delivered their pre-kaizen actions ⇒ All showing behaviours indicating that they are ready and capable of change ⇒ All team members have a basic level of tools skills as needed for the kaizen week	GREEN All team members have cleared their diaries for the kaizen week, completed their actions (such as keeping a time diary as they have gone about BAU in the period leading up to the kaizen) and are excited about the upcoming event. Some natural scepticism at whether the kaizen can deliver all it has promised.
	CSF	**Target – Pre-kaizen**	**Status – Pre-kaizen**
3	Successful kaizen week	⇒ All logistics in place ⇒ All kaizen week success criteria agreed (critical features agreed) ⇒ Data collection complete ⇒ Pre-kaizen data analysis complete	AMBER Some data gaps but all logistics in place including access to other facilitators with business improvement capability
4	A capable Project Manager	⇒ The full project delivery is in place and in use ⇒ The pre-kaizen week targets have been achieved ⇒ Kaizen tools training has been completed ⇒ Kaizen week designed and facilitation organized	GREEN Everything in place as planned. Feedback from team that they are happy with the support and the support mechanisms put in place.
5	Engaged project sponsor and stakeholders	⇒ All active in supporting the methodology ⇒ Active in supporting the resource release and data collection. ⇒ Committed to attending the kaizen daily out briefs, in particular the final day	GREEN Everyone is supportive, if a little sceptical. Resources released and additional funds to take team out for a meal at the end of the kaizen week. Good commitment to attendance at each daily out brief.
6	An organization ready for change	⇒ Communications to pre-kaizen completed and successfully received ⇒ Those impacted are supportive of the change (release data, ask for information, committed to attending at least one kaizen daily out briefs)	AMBER The resistance seen during team selection is still around and needs to be managed. In general other people in BAU team are interested and are pleased with the open invitation to 'come and have a look at what we're doing'.
Decision		**Comments**	
Go		Overall the project pre-kaizen is assessed as AMBER. This is an acceptable level of risk considering where the amber risks are and the mitigation plans in place.	

The conclusion pre-kaizen was that the team was as well prepared as they could be considering that they have never done anything like this before.

As the kaizen event proceeded, all team members were amazed at the way this accelerated way of working engaged them and progressed the project in terms of designing a new purchasing business process.

Table 4-11 Completed Tracking RACI – purchasing business process improvement

Delivery Toolkit – Tracking RACI								
Project: Purchasing business process improvement						**Sponsor:** Barry Jones		
Date: Week 6						**Project Manager** Jo Smith		
Activity tracking								
Activity	**R (responsible)**	**Y/N**	**A (accountable)**	**Y/N**	**C (consulted)**	**Y/N**	**I (informed)**	**Y/N**	**Gap analysis**
Data collection – pre-kaizen	Kaizen team	Yes	Project Manager	Yes	Core team	Yes	Kaizen team	Yes	None Using purchasing expertise appropriately
Data analysis – pre-kaizen	Core team	No	Project Manager	Yes	Kaizen team	Yes	Kaizen team	No	Project Manager did bulk of data analysis due to lack of experience in core team. Informing the wider team pre-kaizen was not appropriate
Kaizen checklist	Project Manager	Yes	Sponsor	Yes	Core team	Yes	Kaizen team and senior BAU team	Yes	None Using project and business change management expertise appropriately
Attend kaizen tool training	Kaizen team	Yes	Project Manager	Yes	Core team	Yes	Sponsor	Yes	Although everyone went through the training it was clear that the baseline capability was lower than expected
Pre-kaizen stakeholder management	Project Manager	Yes	Sponsor	Yes	Core team	Yes	Sponsor	Yes	This took more time than anticipated, and at times consultation with the core team was not possible
Action plan								
Review of 4 key pre-kaizen activities highlighted the novel nature of this project approach and the support that will be needed during the kaizen week in terms of facilitation and data analysis. Additional facilitators have been obtained to support use of the business improvement tools and mitigate capability risks.								

Table 4-12 Completed Project Team Audit Tool – purchasing business process improvement

Delivery Toolkit – Project Team Audit Tool			
Project:	Purchasing business process improvement	**Sponsor:**	Barry Jones
Date:	Week 6	**Project Manager**	Jo Smith
Team checklist			
Check	**Response**	**Comment**	
Is team management appropriate?	Yes	Using the purchasing reps in the core team (Figure 4-16)	
Is team communication appropriate?	Yes	Trying not to overload the team with too much new information	
Is the team working area (office or virtual environment) supportive of success?	Not yet	The team are meeting infrequently and have no team area. This will be solved once we kick off the kaizen week as we will quickly generate a team working area.	
Is team morale supportive of success (how the team behaves, its values)?	Mainly	Team are responding well to a difficult change situation (analyzing their own jobs and to some extent their performance). They are managing queries well from co-workers and facing resistance with maturity.	
Is the team being managed effectively?	Yes	The core team have been fundamental to effective management pre-kaizen as the Project Manager is not involved full time in the purchasing department	
Is the team performing (achieving its goals)?	Mainly	Where data is available, the team have collected it although there is some questioning as to why specific data is necessary. There is a natural challenge to data collection – the team are not used to measuring performance in this way.	
Is there conflict in the team?	No	There is the hint of possible conflict between some team members and an individual co-worker – this needs monitoring	
Is the team working effectively (individual performance, clarity of role and WoW)?	Yes	As well as can be expected pre-kaizen, with only a low level of activity from most team members (as planned)	
Team action plan			
The kaizen team and the core team are working as well as can be expected. They are a team already because all are working in the purchasing department, however, they will only form as a **kaizen team** during the kaizen week. The goal for the Project Manager is to have a high energy day 1 that assists the team to quickly move to being highly performing – this is possible.			

Scope management and delivery

The scope of the project, as defined in the set-up plan, was the improvement of the purchasing business process (Figure 4-19) through using a Lean Six Sigma (LSS) methodology (via a kaizen event). The final deliverables were defined as:

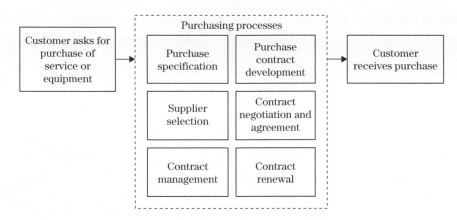

Figure 4-19 Purchasing business process

- A new purchasing business process.
- A set of role descriptions and performance measures.
- A business change plan – changing from current to future ways of working (WoW).

In order to deliver these outcomes, the CSFs needed to be achieved, requiring robust project and business change management. However, the business process design work was mainly conducted during the kaizen week. Success in the latter was seen by the delivery of critical features which the senior stakeholders and business owner expected to see at the end of the project (Figure 4-20). The scope can therefore be considered in two areas: process and content.

Process scope

The team used the work breakdown structure which was developed by first brainstorming all the activities required to achieve each CSF. This was tracked using an activity plan which defined quantity, quality and functionality of each item in the work breakdown structure (activity). The team also managed the dependencies between items. Process scope activities included the design of the kaizen event, the management of stakeholders and communication.

Content scope

The basis for content scope is the CTQ Tree (Figure 4-20), and this supported the development of the design activities which were performed during the kaizen event (Table 4-13). Effectively, the content scope is divided into three areas of design as defined within the CTQ Tree:

- Purchasing process, measures and roles.
- Purchasing organizational design.
- Implementation.

An example of the outputs is collated in Figure 4-21.

The kaizen week delivered the intended scope as defined by the CTQ Tree (Figure 4-20) and gained approval for three levels of business change (Table 4-14):

1. *JDI Phase* – 'Just Do It' immediate changes as soon as the kaizen week finished.
2. *Phase 1* – finalizing the business process and measure changes (design completion and implementation).
3. *Phase 2* – finalizing the reorganization (design, approval and implementation).

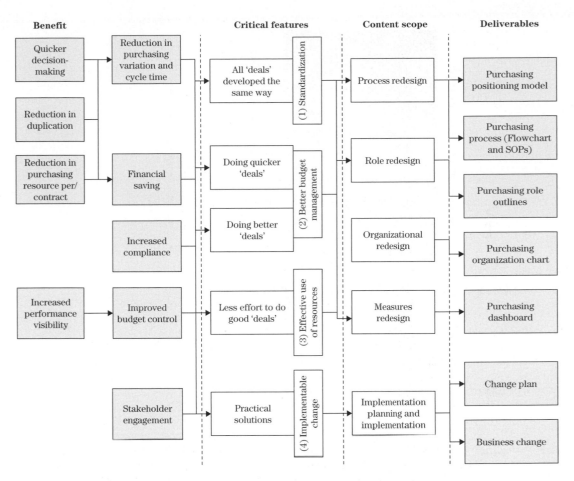

Figure 4-20 CTQ Tree – kaizen success criteria – purchasing business process improvement

Table 4-13 Content scope – purchasing business process improvement

Day	Day 1	Day 2	Day 3	Day 4	Day 5
Activities	⇒ Data mapping ⇒ Process mapping ⇒ Visioning	⇒ Root cause analysis ⇒ Time-value mapping ⇒ Data analysis	⇒ Process design ⇒ Problem elimination ⇒ Benefits specification	⇒ Detailed process design ⇒ Process simulation ⇒ Role design	⇒ Change impact analysis (force field and risk analyses)
Deliverables	⇒ Current state map ⇒ Vision of success ⇒ Purchasing model	⇒ Detailed problem statements ⇒ Waste list ⇒ Current measures	⇒ Purchasing business process (flowchart) ⇒ New measures	⇒ Purchasing business process (SOPs and role outline)	⇒ Purchasing business change plan ⇒ Purchasing organization chart

Day 1 – mapping the current state

- Over 100 issues identified
- Average time for supplier selection = 1 year
- Average time to specify = 3 months

Day 4 – design and testing

- Process simulated for a recent purchasing project and benefits identified in terms of reduced cycletime and resource 'effort'
- Measures and roles developed
- Some areas of the design were implemented – stopping wasteful activities

Day 2 – identifying root causes

- All issues were analyzed using a combination of 'five whys?', fishbone and what-if? techniques

Day 5 – implementation planning

- Developing a business change plan to take the business from the current state to the future state

Day 3 – mapping the future state

- Identification of stages and stage gates
- Linked to a purchasing model with four categories

Post-kaizen

- Monitor implemented changes
- Implement further changes
- Conduct an engagement survey

Figure 4-21 Kaizen photos – purchasing business process improvement

Table 4-14 Completed Scope Tracker – purchasing business process improvement

Delivery Toolkit – Scope Tracker

| Project: | Purchasing business process improvement | | | Sponsor: | Barry Jones | | |
| Date: | Week 6 | | | Project Manager: | Jo Smith | | |

Deliverable	Change?	Scope	Impact?	Critical feature	Impact	Benefit criteria	Impact
Agreed output from the project	*Add, delete, modify*	*Agreed scope to deliver the CTQ feature*	*Impact of change to deliverable on scope*	*A CTQ feature*	*Impact of change on CTQ feature*	*Insert organizational or project benefit*	*Impact of CTQ feature change on benefit*
Purchasing model	None	Process redesign	None	(1) Standardization; (2) Better budget management	None	(a) Reduction in purchasing variation and cycle time; (b) Financial savings; (c) Reduction in purchasing resource/contract; (d) Increased compliance	None
Purchasing flowchart	None	Process redesign	None	(1) Standardization	None		None
Purchasing SOPs	**Modify**	Process redesign	Reduction in detail completed during kaizen	(1) Standardization; (2) Better budget management; (3) Effective use of resources	None – this is a timing change		Benefit reduced until phase 1 completion
Purchasing role outlines	None	Role redesign	None	(1) Standardization; (2) Better budget management; (3) Effective use of resources	None	(e) Reduction in duplication; (f) Quicker decision-making	None
Purchasing dashboard	None	Measures redesign	None	(1) Standardization; (2) Better budget management	None	(g) Improved control of budget; (h) Increased performance visibility	None
Business change plan	**Modify**	Implementation planning and implementation	Only a third in kaizen scope	(1) Standardization; (2) Better budget control; (3) Effective use of resources	None – this is a timing change	(i) Stakeholder engagement in all changes	Benefit reduced until phase 2 completion
Organization chart	**Modify**	Organizational redesign	Removed from kaizen scope	(3) Better budget management	None – this is a timing change	(g) Improved control of budget	Benefit reduced until phase 2 completion

Scope change summary

All scope changes were accepted although it was recognized that these changes would reduce some of the immediate impact of the kaizen. However, due to other organizational changes going on in the business it was the right decision to not fully redesign the purchasing department within the kaizen. In any event, the reduction of scope gave the team more time to spend on the main areas of scope – the process redesign.

The four key concepts underpinning the scope of the business change were delivered and are discussed below:

Purchasing positioning

The segmentation of purchasing (and then supplier associated with that purchase) is now according to risk and purchase level. This allows prioritization of purchasing resource and appropriate value-add management of suppliers. For example, a supplier who is delivering a highly critical, high cost product or service would be managed quite differently from one providing simple, low cost consumables.

Standardization

The generation of a standard set of processes (with associated documents and measures) based on best practice in the network, with links to purchasing positioning so as to be 'standardization with common sense'.

Value stream approach

This approach refers to the generation of a flow of value for the business from the decision to 'buy,' through to contract implementation and renewal until the decision 'not to buy'. This is the foundation for the business processes, the culture and the organizational structure/roles, *but* does not define any of these three without consideration of the other key concepts. A value stream approach aims to:

- Minimize transitions and handovers, eliminating functional silos, concentrating on the decrease in cycle time and inventory, the increase in purchasing effectiveness and the reduction in variation.
- Impact the organizational design so that roles cross over traditional purchasing boundaries in the company.

Measures

The generation of appropriate measures driven by the three previous concepts:

- The purchasing positioning drives purchasing category measures, which drive appropriate behaviours in each category.
- Standardization drives the development of a dashboard approach to focus on key measures.
- The value stream drives overall measures of cycle time, inventory and throughput (in terms of getting contracts implemented and realizing benefits) and also variation.
- Measures were divided into those measuring the process and those measuring value generation.

Conclusions

Overall, the project delivery took 9 weeks as scheduled; 6 weeks pre-kaizen and 3 weeks for the kaizen and post-kaizen scope. At the close-out review, the following achievements were acknowledged:

- Reduced cycle time (from decision to 'buy' to an implemented contract) – potentially from 12–18 months to 3 months.
- Reduced cycle-time variation – potentially from 'random' to ±2 standard deviations.

- Reduced 'work in progress' (purchasing contracts not yet completed).
- Reduced amount of purchasing resource needed due to increased effectiveness and efficiency (as measured by reduction in total resource/contract).
- Savings to the business released earlier – an implemented contract may save money for the business. If this can be achieved earlier, then each day earlier represents financial savings for the business.
- Improved purchasing team development due to all having clear roles and associated development targets.

The kaizen was a very dynamic and inspirational way to design and implement change, but it required strong leadership support and effective communication to those impacted by the changes. The Project Manager understood the risks of this approach and ensured that they were managed within the control plan.

In order to sustain the business benefits, the purchasing team need to be engaged in the change, and Figure 4-22 demonstrates that this is growing.

Status of engagement

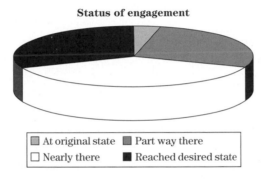

Figure 4-22 Team engagement – post-project – purchasing business process improvement

In addition, the team confirmed the following:

- The project outcomes have changed personal decision-making on what it is appropriate to spend time on – this is generating a consistent approach across the UK.
- The value-stream approach is driving positive behaviours which support value for the business – reducing duplication of effort.
- It's great to have some new measures.
- Expect 'noise' – this is a change project and some noise is expected – not everyone went through the 'kaizen experience'.

Key points

The aim of using this case study was to demonstrate that:

- Delivery of the set-up plan was fundamental to success because people and scope were at the heart of this project.
- Getting the right mix of capabilities within the Project Team is critical to project success.
- A project can take many different routes, and a good Project Manager will define the direction and the key decision points along the journey – this is the value of a project roadmap.

Troubleshooting set-up plan delivery

Table 4-15 is a list of common issues associated with set-up plan delivery.

Table 4-15 Troubleshooting set-up plan delivery

Typical symptoms	Example root causes	Example solutions
1. Not following the set-up plan	Project Manager has not engaged with the Project Team	Ensure that the project has been kicked off appropriately
2. Poor team performance	Inappropriate Project Team selection (linked to poor planning) – team lacks capability	Review the capabilities required and work with sponsor to select and get release of appropriate people for the team
	Poor team or individual management – Project Manager may not have the people skills	Project Director or line manager to mentor the Project Manager with regard to team management
	Poor or no team engagement with the project goals	Ensure that a team engagement plan identifies the right way to engage the team and keep them engaged. For example: ➠ Communications ➠ Kick-off ➠ Measures and reward
	Inappropriate communications	Ensure that communications plan includes the Project Team as a distinct stakeholder group
	Team of one	Ensure that the Project Team is appropriately resourced with each member having a clearly defined role, responsibility and accountability (RACI, resource-loaded schedule)
	Too many chiefs	
	Too many Indians	
	Part-time team members	
	Gaining the wrong team members (being given team members to get them out of other people's way)	Work with the sponsor to position that this is the wrong person for the project
		Identify a low-risk activity where the team member could deliver value
3. Team conflict and/or conflict with the Project Manager	Breakdown of authority in the project	Consider changing the Project Manager – highly unlikely that authority can be regained by an external intervention
	Cultural or personality clashes	Consider the individual situation and individuals involved. First response may be mentoring, but ultimate response may be to remove one or both individuals from the project.

(Continued)

Table 4-15 (Continued)

Typical symptoms	Example root causes	Example solutions
4. Team instability – lots of changes	Losing team members – business will not commit permanent members to the project	Use sponsor to help business to understand the need and benefit of permanent team membership
	Losing team members – team members request to leave	Ensure a formal exit review so that reasons are fully understood and mitigating actions can be taken to prevent impact on wider team
	Gaining team members	Ensure appropriate new member induction and structured team-building events. Consider communication plan to link in with key phases in team growth.
5. Finance and project have different views on total spend to date	There is no link between the project and the financial management within the company	Ensure that financial controls align with business needs by developing a working relationship with accountants
6. Pressure from financial group	Dealing with year-end issues – meeting cash flow targets (spent too little or too much)	Ensure that an effective relationship between the project and business accountants so that funds are spent at an agreed rate, in line with business year-end targets and at times to be of most benefit to the organization as a whole
	Dealing with year-end issues – meeting business targets	
	Currency volatility	
7. Scope keeps changing	Not defined well enough at the beginning	Stop and conduct a full scope review to ensure that scope is linked to critical needs/benefits
	You were doing the wrong project in the first place	
	Customer changes scope	Review required changes versus the original business case. Conduct a value review exercise to determine if the change will deliver additional value to the business.
	Management changes scope	
	Supplier changes scope	
	Business is changing	Communicate why the change has occurred – to ensure that the scope continues to align with the business need and delivers appropriate value
8. Scope creep	The Project Team are overengineering the project	Ensure that the Project Team understand what is critical to quality and only change where there is clearly identifiable value
	Scope management is not robust	Ensure that there are appropriate deliverables tracking mechanisms in place that link scope to benefits

Handy hints

Project management is all about people management

Be sure to spend time considering team culture and team motivation.

Value management needs to be integrated into everyone's way of working if it is to be successful

Understanding and delivering value can be confusing especially with a large team, a complex project or a customer group with varying expectations – have processes in place for the cascade of value management so that scope delivered is linked to the benefits required.

Manage organizational resources wisely

This might seem obvious, but any Project Manager needs to realize that when organizational resources (e.g., money and people) are allocated to his project, they cannot be used elsewhere. In effect any misuse by you represents a waste of resource efficiency to the business which can never be recouped.

No matter how simple the project or how similar it may be to a previous project, define a project roadmap

A roadmap is a communication tool for everyone involved in a project. It tells us where we are going and what we have to do to move through successive stages of the journey.

Don't ignore finance management – there are organizational and legal implications

Organizations usually have very strict rules and guidelines on how money can be spent. This is usually linked to their own business governance as well as accounting law. Find out the rules and make sure that the team understand them.

Delivering a project successfully needs people – don't ignore the team

A key part of maintaining engagement with the Project Team is involving them in appropriate decisions and in empowering them clearly. However, situations can change, and be sure to communicate effectively with your team when decision-making processes, authority limits and overall goals alter and *why*.

And finally . . .

Effective set-up plan delivery:

- Maintains a link between the people within the project and the value that they have to deliver.

- Supports business plan delivery, ensuring that the business benefits will be enabled.

- Ensures that the control plan is appropriate and capable of supporting project success.

- Is built on effective Project Team management and effective value delivery.

5 Control plan delivery

In the context of project delivery, the control plan is the way we control the project. It is based on the approved project business case which explains 'how' the project needs to be delivered to meet the business needs. The robust delivery of the control plan relies on the Project Manager and Project Team working together to ensure that the project manages uncertainty. The two concepts fundamental to control plan delivery are:

- *The control loop* – the monitoring of project activities against plan in order to develop a forecast to completion.
- *Uncertainty management* – the review of, and reaction, to risks (opportunities or threats) to project completion.

In order to manage these concepts, the Project Manager needs to use an appropriate set of tracking tools which support a focused, but flexible, approach to delivery of the control plan.

What is a project control plan?

As defined in detail in *Real Project Planning* (Melton, 2008), a project control plan is that part of a Project Delivery Plan (PDP) that is the formal articulation of *how* the project will be managed and controlled, so that the business is assured of the certainty of outcome with respect to the original business case. It covers the following four planning and delivery themes:

- Risk management.
- Contract and supplier management.
- Project controls.
- Project review.

How to manage delivery of a project control plan

To track the delivery of the project control plan, a Project Manager must actively manage the control strategy by behaving both as a 'control freak' and as devil's advocate (Figure 5-1).

The Project Manager has to consider how best to manage the agreed control strategy considering all 'hard' and 'soft' aspects. Although the key relationship in delivering the control plan is with the Project Team, the Project Manager also needs to consider the wider stakeholder group, particularly the customer and end-users who can significantly impact change management.

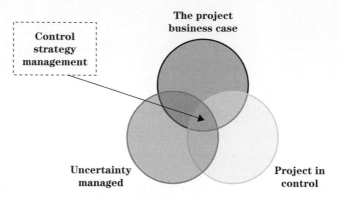

Figure 5-1 Control strategy management

The Project Manager as 'control freak'

The traditional role of the Project Manager has been to control all the project activities so that the project is delivered as planned. This means delivering all project objectives and in particular those associated with cost, time and change. A typical model of control (Figure 5-2) demonstrates how

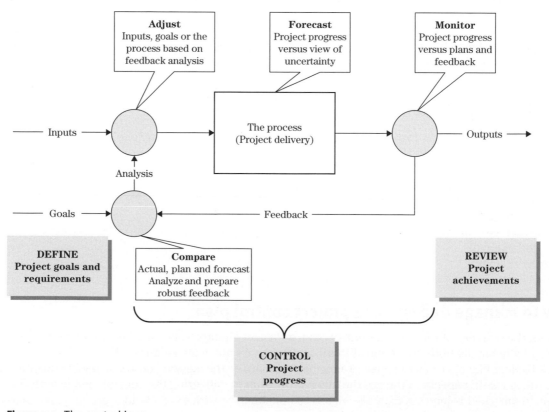

Figure 5-2 The control loop

control is a function of having goals, attempting to meet them and then, based on a realistic forecast, adjusting delivery to achieve project success.

Without continually forecasting the end point, a Project Manager is not controlling but merely monitoring, and a good Project Manager needs to be a control freak (continually looking at progress in the context of the uncertainties within the project and business environment).

The Project Manager as devil's advocate

A key role which the Project Manager must play is that of devil's advocate: debating with the Project Team from a view he or she doesn't necessarily hold, but which challenges the team view (thus assuring the Project Manager of its validity). In effect, sometimes a Project Manager has to argue with the team for the sake of argument. This is about testing assumptions and managing uncertainty so that progress is maintained and a successful outcome forecasted.

Risk and issue management

Whether a control freak or devil's advocate, a Project Manager needs to maintain control and a crucial part of control is managing uncertainty – risks and issues:

- A risk is something which could occur but has not yet happened.
- An issue is a risk which has occurred.

Both need to be managed proactively during project delivery according to the risk management plan (Figure 5-3).

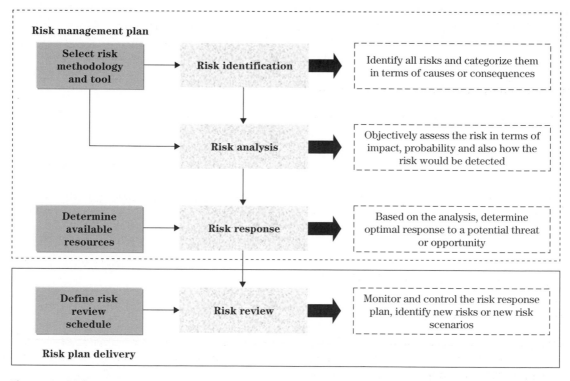

Figure 5-3 Risk management process

Risk plan delivery

During the project planning phase, the risk management plan is developed; during delivery, it is therefore implemented. Risk plan delivery should:

- Consider both opportunities and threats as project delivery progresses. Both are elements of uncertainty and need to be managed. Opportunities need to be maximized, and threats need to be eliminated or minimized.
- Include the development of a structured risk response management process appropriate to the project and the risk management tools and techniques used during the planning stage. The

risk baseline is in the PDP and needs to be used as the foundation for the management of uncertainty.

The main focus in delivering the risk plan is the review of its progress as the project is being delivered. This relies on:

- Having used the most appropriate risk tools and techniques during planning to baseline the project risk distributions.
- Having risk reviews as an integral part of the project delivery process – with links to reporting, communication and action planning.

Risk management

A risk plan that is developed but never used to control a project is a waste of time and introduces a major risk to the project in its own right. A project is based on uncertainty, and this needs to be actively managed.

The management of risk needs to be in an appropriate place in the project organization. Risk needs to be managed by those best placed to manage it. For example:

- *Overall project risks* – The Project Manager is best placed to manage these risks based on appropriate information/control data from his team.
- *Specific work package risks* – The Project Manager should delegate specific technical risk to the parts of his team that have the knowledge and skill to develop specific mitigation plans to deal with the risks.
- *Specific supplier work package risks* – The contract with a supplier may have been set-up so that they are responsible for specific risk areas where they have the technical ability, and it is fundamental to the delivery of the work package they are contracted to deliver.

Risk management should be cascaded throughout a Project Team. Everyone has a part to play in terms of managing, mitigating and communicating risks.

Risk tools and techniques

Assuming appropriate risk tool selection during planning, the chosen tool must be used during delivery and the results appropriately interpreted. Table 5-1 describes the ways risk tools can be used and the outputs interpreted to support project forecasting.

The key to risk management during project delivery is the interpretation of the results of each risk review and then the use of that data in forecasting.

Communicating risk status

In selecting the risk management tools and techniques, a key consideration is how the risk data will be used in project communications. As with any communication, a risk message needs to be customized for the particular audience so that they understand what they need to do as a result of receiving that message. For example:

- Communicating risk status to the Project Team needs to be at a level that they can use to cascade into sub-project teams. The Risk Table and Matrix (Figure 5-4) can be useful as it collates a lot

Table 5-1 Example – risk tool use during delivery

Risk tool	Description	Output interpretation
1. Risk Table and Matrix (Figure 5-4)	To identify risks to achievement of project CSFs and, through prioritization, the identification of appropriate mitigation and contingency plans. Can also be used on a WBS and categorization other than CSF.	⇒ Identifies CSF or other categories within the project scope where risks are highest and can forecast likely mitigation and therefore project outcome. ⇒ Supports resource management through prioritization – the total red, amber or green risks is an indication of the resource level need for risk management
2. Critical Path of Risks Table (Figure 5–5)	This tool relies on using the risk table and matrix and is effectively an 'add on' visual tool to show a risk profile versus the CSFs	⇒ Can give a good high-level view of the current project situation and the level of risk to a successful project outcome. ⇒ The data can be used to populate a risk profile. (Figure 5-6) ⇒ Can be used to forecast which CSFs will/will not be achieved.
3. FMEA (Failure Mode Effect Analysis)	This tool is a methodology focused on the elimination of defects based on a risk priority score	⇒ To assess and correct individual parts of a process, system or product and therefore to forecast its likely outcome. ⇒ Useful for managing the most critical defects within a sub-system and useful for cascading risk management. ⇒ The trend in risk priority scores can be evaluated and used for forecasting.
4. Risk Profile (Figure 5-6 and Table 5-2)	This tool is used to show project risk trends during project delivery	⇒ The risk trend can be evaluated and outcomes forecast as a result. ⇒ The risk profile can be for the whole project or a specific part and the format of the profile depends on the risk technique used as the basis.
5. HACCP (page 145)	This tool identifies the most critical areas of a project that require controlling in order to ensure project success	⇒ The critical risk areas will be seen as in or out of control. ⇒ If a critical control area is in control then the control mechanism is working and the risk is unlikely to impact the project outcome. The converse is also true.
6. Project SWOT Table (Melton, 2008)	To identify and analyze project strengths, weaknesses, opportunities and threats so that an appropriate action plan can be put in place	⇒ Can give a good overall review of the current project situation. ⇒ The progress of the SWOT action plan can be used to forecast how the SWOT profile will change.
7. Project Scenario Tool (Melton, 2008)	A method to identify specific project scenarios considering specific parameters	⇒ Can be used to forecast how the project would cope with different future scenarios. ⇒ The scenarios need to be based on realistic and structured analysis of possible future events.

(Continued)

Table 5-1 (Continued)

Risk tool	Description	Output interpretation
8. Risk flowcharts	This tool is based around a decision tree methodology asking risk related questions and determining action based on the response	⇒ During project delivery this tool can be used to quickly identify high risk areas and critical actions for risk reduction. ⇒ This tool is useful to use in project audit or health check situations when forecasting whether a project or sub-project is in control and capable of delivering a successful outcome.
9. Risk checklists	This tool is based around a keyword methodology asking risk related questions against key risk areas	⇒ During project delivery this tool can be used to quickly identify high risk areas and critical actions for risk reduction. ⇒ This tool is useful to use in project audit or health check situations when forecasting whether a project or sub-project is in control and capable of delivering a successful outcome.
10. Scope/cost or schedule risk assessment	This tool uses the basic risk process against a specific consequence category (scope, cost or schedule)	⇒ During project delivery these tools are used to track specific risks and then forecast the project outcome in that area. ⇒ For example, a schedule risk assessment allows the Project Manager to forecast the likely schedule outcome. ⇒ For example, a cost risk assessment allows the Project Manager to forecast the likely achievement of the cost plan.

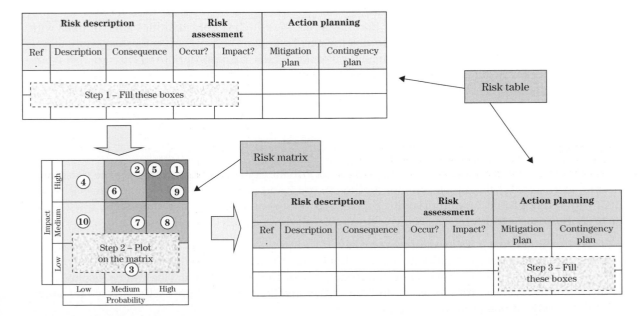

Figure 5-4 Risk Table and Matrix

of risk data and then displays it visually. Each sub-project team can generate their own matrix and track how the risks are moving from red to amber to green as mitigation plans are delivered effectively.

▶ Communicating risk status to the sponsor or customer needs to be at a high level and be capable of communicating the overall probability of project success. The Critical Path of Risks (Figure 5-5) is a simple visual tool which can convey the current risk status of a project including the probability of a successful outcome.

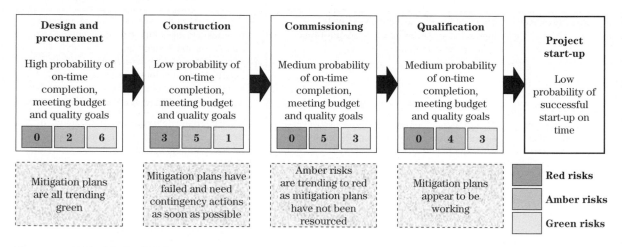

Figure 5-5 Critical Path of Risks

▶ At a very detailed level, a Project Team may use a risk log which simply identifies a list of risks with assigned mitigation responsibilities. The risk log may be kept as a 'live' tool on a project notice board so that team members can prioritize their tasks. Whilst it is appropriate to cascade risk where it can be effectively managed and controlled, the process needs to ensure that critical risks which can impact the project outcome are identified and communicated. Risk status needs to cascade both up and down.

Risk profiles

Communicating current risk status is an important part of managing risks and delivering the risk plan. However, as risk status can move, another method to track and communicate project progress is to use some form of risk profile. Most tools deliver a summary status on risk level, and this can be tracked over time. For example:

▶ Track the FMEA risk priority scores over time.
▶ Track the resulting project risk status generated from a Critical Path of Risks over time.
▶ Track the total number of risks generated from the Risk Table and Matrix. This can be displayed in a bar chart format (histogram) segmenting red, amber and green risks (Figure 5-6). This would show the risk profile at one point in time only.

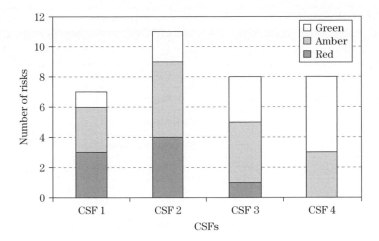

Figure 5-6 Example risk profile

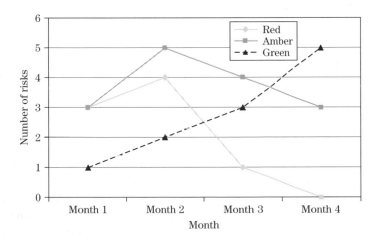

Figure 5-7 Tracking risk profile

➤ Track the total number of red, amber and green risks generated from the Risk Table and Matrix (Figure 5-7). This produces a history of the risk profile over time.

Risk profiles can display a project trend that could be missed by other, more traditional project control tools. Apart from using them to review the whole project, risk profiles can be used within a sub-project to trend sub-project risks.

Tool: Project Risk Profile

The aim of this tool (Table 5-2) is to allow analysis of the overall trends in the risk profile of a project or part of a project.

Table 5-2 Project Risk Profile explained

Delivery Toolkit – Project Risk Profile						
Project:	*<insert project title>*		**Project Manager:**		*<insert name>*	
Date:	*<insert date>*		**Page:**		*1 of 1*	
Risk profile trend						
Risk area	**Milestone review dates**					
	<insert date>	*<insert date>*	*<insert date>*	*<insert date>*	*<insert date>*	*<insert date>*
<Insert CSF or risk category>	*<insert risk rating>*	*<insert risk rating>*	*<insert risk rating>*	*<insert risk rating>*	*<insert risk rating>*	*<insert risk rating>*
Project risk summary						
<insert summary comments on the status of the risk profile and the trends seen>						

Risk profile trend

Risk area

Select the method to categorize the critical areas of the project or sub-project. This will usually have been done in some way during the planning stage and is typically generated from:

- The project path of critical success factors (CSFs).
- The categories of risk consequences; for example, cost, schedule, scope.
- The project or sub-project failure modes.

There needs to be a more detailed risk assessment behind whichever risk area is chosen. For example:

- If the Risk Table and Matrix and Critical Path of Risks are used with the path of CSFs, then the CSF risk ratings would be used. Typically these are RAG ratings: red, amber or green.
- If the FMEA is used, then the risk rating would be expressed as the risk priority number. If the FMEA is comprehensive, then the profile might only track the highest scoring risks. This is appropriate as long as other risks are added to the profile if their scores increase.

Milestone review dates

The review dates would be linked to dates when the detailed risk review is taking place. At each milestone date, the detailed risk review would be analyzed, and a RAG rating (or risk score) determined for each risk area.

Project risk summary

Based on the trends in risk profile, the Project Manager may choose to highlight where further review and/or mitigating action needs to be taken. Typically the profile will be used to forecast the outcome of each risk area and how this impacts the overall project outcome.

The risk data can be converted into a histogram (Figure 5-6) or a scatter diagram (Figure 5-7) to support these trend reviews and outcome forecasting.

Short case study

An engineering institution wanted to raise its profile in the industry. In order to do this, it recognized that there needed to be some change in the way it operated and so commenced a business change project with the goal of 'reputation enhancement'.

A series of CSFs were determined at the start of the project along with their risk rating. The Project Manager believed that the project was relatively low risk and therefore did not review the risks regularly. However, in Month 4 a major issue with the performance of the external PR company instigated a full project risk review. It was a surprise that not only had CSF 2 moved to a higher risk status but also CSFs 1 and 5 were trending that way (Table 5-3). Such a negative trend was highlighted as a serious issue. Mitigation actions were put in place prioritizing CSF 2 (as resources were limited) and, as seen by Month 8, the risk profile was more positive.

Table 5-3 Example Project Risk Profile

Delivery Toolkit – Project Risk Profile						
Project: Reputation Enhancement			**Sponsor:**	Tom Smith		
Date: Month 8, year 1			**Project Manager:**	Jerry Jones		
Risk profile trend						
Risk area	**Milestone review dates**					
	Project start	**M2**	**M4**	**M6**	**M7**	**M8**
CSF 1-Strategy clear	Green	Amber	Amber	Amber	Amber	Green
CSF 2-Public relations successful	Green	Amber	Red	Red	Amber	Amber
CSF 3-Successfully change WoW	Amber	Amber	Amber	Amber	Amber	Amber
CSF 4-Stakeholders engaged	Amber	Amber	Amber	Green	Green	Green
CSF 5-Resource capacity and capability	Green	Green	Amber	Amber	Red	Amber
CSF 6-Institutional governance	Green	Green	Green	Green	Green	Green
Project risk summary						
The trends are now moving in the right direction showing that the mitigation actions started in Month 6 are working. The static nature of CSF 3 is a growing concern as it is getting harder to forecast. Need to have a review of the risks and activities associated with this CSF.						

The Project Manager was certain that the risk profile approach had presented a more strategic view of risks to the senior stakeholders and enabled the release of resources which allowed mitigation plans to be actioned for most red and amber risks.

Hazard Analysis and Critical Control Point (HACCP) methodology

HACCP is a risk methodology which has been extensively used in the food industry to identify hazards which are potential food safety issues. The technique can be used in a project context because it is based on the identification of critical areas which are impacted by hazards.

A generic HACCP process

HACCP is a systematic process and has a defined route to identify and then manage hazards. Within a project context, the specific process used in the food industry can be adapted (Figure 5-8).

The process relies on the six steps shown in Figure 5-8, and therefore the following definitions.

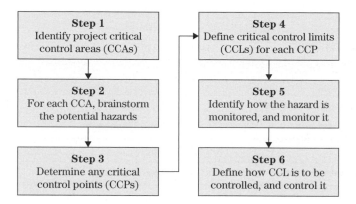

Figure 5-8 Project HACCP process

Project HACCP definitions

The project HACCP process (Figure 5-8) relies on the use of two key project management principles:

➤ The definition of a project critical path of success.
➤ The definition of a project roadmap.

Control area

Any stage within the project roadmap which can impact the achievement of a CSF.

Critical Control Area (CCA)

A Critical Control Area (CCA) is a situation within a control area where control needs to be applied to prevent or eliminate a project hazard, or reduce it to an acceptable level.

Hazard

A project situation that is reasonably likely to cause the failure of a project CSF in the absence of its control.

Control point

A situation within a project control area where cost, schedule, scope or benefits can be controlled.

Critical Control Point (CCP)

A Critical Control Point (CCP) is a situation within a project CCA where control can be applied and is essential to prevent or eliminate a project hazard, or reduce it to an acceptable level. To determine whether a control point is a CCP, a project CCP decision tree is used (Figure 5-9). This contains a sequence of questions which test if a control point is actually a CCP.

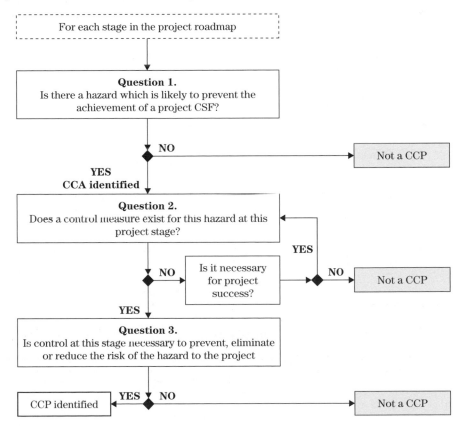

Figure 5-9 Project CCP decision tree

Critical Control Limit (CCL)

A Critical Control Limit (CCL) is a maximum, and/or minimum value to which a project parameter must be controlled at a CCP to prevent, eliminate, or reduce to an acceptable level, the occurrence of a project hazard.

HACCP management during project delivery

In order for the process to be effective, the HACCP results should be reviewed frequently during project delivery to establish any changes or trends in CCLs. For this to be robust, the hazard identification and CCL control processes need to be effective and effectively communicated. Realistically there should be a limited number of CCPs:

➤ Too few – and it may be that the CSFs for the project have not been adequately understood.
➤ Too many – and the project risk profile may become unmanageable or the CCP definition is flawed.

A typical set of templates to report the project HACCP process during delivery are contained in the Project HACCP Tool (Tables 5-4 and 5-5).

Tool: Project HACCP (parts a and b)

The project HACCP analysis is a 2 part tool. It aims to provide the Project Team with a technique to identify, manage and control specific, detailed areas of project activity. It relies on:

➤ The identification of CCAs and within these CCPs.
➤ The identification of hazards likely to impact these.
➤ The management of CCLs through definition of a method to identify and control the hazard.

The 2 parts of this tool are:

➤ Part a – Project HACCP Evaluation
➤ Part b – Project HACCP Hazard Analysis and Control

Part a – Project HACCP Evaluation

The Project HACCP Evaluation Tool (Table 5-4) is used to document the first step of the project HACCP process (Figure 5-8). It documents the identification of the Critical Control Areas in the project.

Table 5-4 Project HACCP Evaluation explained

Delivery Toolkit – Project HACCP Evaluation					
Project: <insert project title>			**Project Manager:** <insert name>		
Date: <insert date>			**Page:** 1 of 1		
Project vision					
<insert a short summary of the overall project vision – what change does the project intend to deliver?>					
Critical Control Area (CCA) analysis					
Project roadmap stage	**CSF1**	**CSF2**	**CSF3**	**CSF4**	**CSFn**
	<insert description of CSF1>	<insert description of CSF2>	<insert description of CSF3>	<insert description of CSF4>	<insert description of CSFn>
<insert project roadmap step>	<insert an 'X' and a sequential number or leave blank>	<insert an 'X' and a sequential number or leave blank>	<insert an 'X' and a sequential number or leave blank>	<insert an 'X' and a sequential number or leave blank>	<insert an 'X' and a sequential number or leave blank>

Table 5-5 Project HACCP Hazard Analysis and Control tool explained

Delivery Toolkit – Project HACCP Hazard Analysis and Control							
Project:		*<insert project title>*			**Project Manager:**		*<insert name>*
Date:		*<insert date>*			**Page:**		*1 of 1*
Hazard	**CCA number**	**CCP (Y/N)**	**Critical Control Limit (CCL)**			**Hazard identification**	**Control mechanism**
			Measure	**Target**	**Actual**		
<insert a description of the hazard developed from the evaluation>	*<insert the number(s) of critical control area(s) affected by this hazard>*	*<confirm whether this hazard is a critical control point or not>*	*<insert the measure and units>*	*<insert the target value>*	*<insert the actual value>*	*<insert the control mechanism to be applied>*	*<insert description of how the control action will be applied>*
Project HACCP summary							
<insert summary comments on the status of the HACCP and impact on the project outcome>							

Project vision

It is useful to have a clear articulation of the project vision of success. Achievement of the project CSFs will achieve this.

Project roadmap stage

The project roadmap and associated stages and stage gates will have been defined within the PDP, and so this is simply a listing of those stages.

Critical success factor

The project CSFs will have been defined within the PDP, and so this is a listing of those CSFs. It is often the case that when a detailed hazard analysis is completed, a review or rewording of CSFs is necessary. In extreme cases, additional CSFs may be defined and very occasionally a CSF may be deleted or integrated into another.

Critical control area analysis

Each combination of roadmap stage and CSF is a potential critical control area (CCA). By asking question 1 of the CCP decision tree (Figure 5-9), the CCAs can be defined. Each should then be numbered for reference in the next stage (part b) of the project HACCP process. Once the CCAs have been identified, the HACCP can continue onto part b using the Project HACCP Hazard Analysis and Control tool.

Part b – Project HACCP Hazard Evaluation and Control

This tool (Table 5-5) documents Steps 2 to 6 of the project HACCP process (Figure 5-8). It identified the way the hazards will be controlled.

Hazard

Brainstorm and insert the potential hazards which could occur within each CCA.

CCA Number

Collate common hazards across a number of CCAs, and the CCA numbers should then be inserted here.

Critical control points (Y/N)

By using questions 2 and 3 of the CCP decision tree (Figure 5-9), the CCPs can be defined for each potential hazard. A 'yes' (Y) or 'no' (N) should be inserted, although for reviews it would be usual to omit the non-CCPs.

Critical control limit

As previously defined, these are limits within which a hazardous situation must be controlled in order to safeguard project success. For each CCL, there should be a defined measure with units and a target range, minimum and/or maximum. Then each time the HACCP is reviewed, the actual measure for each CCL should be reported.

Hazard identification

Describe the project control process which will identify the hazard, should it occur.

Control mechanism

Identify project control process which will be used to bring the CCL back into the accepted range. This is effectively the preventative action should the hazard occur.

Project HACCP summary

Each time the HACCP is reviewed, the Project Manager should interpret the results of the analysis and summarize the probable outcome for the project.

An example of the use of this tool is shown in the case study in Chapter 7.

Managing issues

During project delivery, some of the risks identified during planning may occur. Once a risk has occurred, it is called a project issue. Project issues can also occur due to unforeseen and therefore unmanaged risks. Once an issue occurs, an automatic reaction is to 'fight the fire', but this is not always necessary nor appropriate in terms of controlling the project and protecting the successful outcome.

The key goals in issue management are:

➡ To fight the 'right' fires.
➡ To optimize the use of a limited pool of resources.

Therefore a Project Manager will usually have a process to log, respond and track project issues (Table 5-6).

Table 5-6 Issue management

Issue type	Immediate response	Response goal
Critical risk A risk which has the potential to impact the overall project outcome	As the impact on the project is critical, it is highly likely that a contingency plan would have been developed during the risk management process. The immediate response is therefore to commence the contingency plan.	To minimize the impact of the issue on the overall project outcome
Non critical risk A risk which impacts a non-critical aspect of the project	During the risk management process a non critical risk may still have a contingency plan ready to be used if the risk was highly likely to occur. In that case the contingency plan may be commenced if the project has the resources available and the plan would prevent future reoccurrence. If no plan is ready it is likely that the only response is an impact review.	To ensure that the impact is still not critical or near critical. To use resources wisely taking into account other project priorities.
Unforeseen critical issue An issue which impacts the overall project outcome	The immediate development of a contingency plan – potentially moving resources from other less critical activities	To minimize the impact of the issue on the overall project outcome and to minimize impact on dependent project activities
Unforeseen non-critical issue An issue which impacts non-critical aspect of the project	Conduct an impact review. If there are resources available then consider what contingent actions would prevent reoccurrence.	To ensure that the impact is not critical or near critical. To use resources wisely taking into account other project priorities.

Short case study

A product launch project was suddenly facing a critical issue because no supplier was able to provide the main raw material. A recent audit had identified that the supplier was not able to maintain the quality to the required specification. This had not been identified as a potential risk. The Project Manager called the team for an action-planning session, and it was agreed to work closely with the supplier in order to improve their quality. This took significant resource.

During this time, other issues occurred: team members left, the manufacturing process testing delivered 40% defects, the regulatory process stalled and market forecasts altered significantly. Because the team was so focused on the supplier issue management, they missed the critical issues appearing in other parts of the project. Finally a risk review highlighted that the risk profile had converted into a very negative issue profile. At this stage, resources were reallocated so that all critical contingent plans could be progressed.

The product launch was delayed but was eventually a success for the organization as product sales exceeded expectations and the project had developed a robust expandable supply chain.

Tool: Issue Action Manager

The aim of this tool (Table 5-7) is to log, analyze, respond to and track the progress of project issues so that resources are prioritized and the right contingent plans are progressed effectively.

Issue log

Issue number and issue owner

Each issue is given a unique number for future reference, and each should have an owner, someone who is responsible for developing and delivering the action plan and reporting current status.

Table 5-7 Issue Action Manager explained

Delivery Toolkit – Issue Action Manager							
Project:	*<insert project title>*			**Project Manager:**		*<insert name>*	
Date:	*<insert date>*			**Page:**		*1 of 1*	
Issue log							
Issue number	**Issue description**	**Issue owner**	**Issue status**	**Issue action plan**		**Action owner**	**Action status**
<insert reference number>	*<insert an accurate description of the issue in terms of the impact>*	*<insert name>*	*<insert issue status>*	*<insert specific actions to address the issue impact>*		*<insert name>*	*<insert action status>*
Issue action summary							
<insert summary comments on the status of the project issues and impact on the project outcome>							

Issue description

A clear statement of the issue should be developed. This should be written in terms of the impact on the project outcome. In addition, some analysis should have been completed, and the root cause identified.

Issue status

Each issue should be designated as critical or non-critical and live or closed.

Issue action plan

Depending on the status, a series of action activities should be detailed. This includes any decision on inactivity.

Action owner

Each individual action should be allocated an owner who has access to the required resources (people, assets, funding).

Action status

Each action should be tracked in terms of completion to plan and impact on project outcome. In other words, is the action plan working to eliminate any adverse impact on critical areas of the project?

Issue action summary

A Project Manager should be regularly reviewing the issue log and analyzing current status in terms of the cumulative impact on the project. A part of this review is the allocation or reallocation of resources.

Contract and supplier management

Almost all projects require external resources for their successful completion (people, goods, services). The contract and supplier strategy should have been defined as part of the project planning phase as outlined in *Real Project Planning* (Melton, 2007). Within project delivery, the objective is to ensure that expectations are managed and control maintained as this strategy is delivered.

Contract management

A contract is a formal definition of the obligations of procurer and supplier within a project. Less formally, it helps frame and define management style and relationship between the parties.

Many different forms of contract exist, but they typically fall into the following two categories:

- *Fixed-price, fixed-scope* – Otherwise known as 'lump-sum turnkey', this form of contract has a defined scope and a defined cost. The obvious benefits of this approach are that the project scope and costs can be well defined. The challenge with managing this form of contract is largely in ensuring the scope is actually fixed and that the basis of procurement will not drive inappropriate behaviours (fixed price, but endless variations).
- *Reimbursable* – This may be against a scope or defined programme of work or a target price. The benefits of this approach are when the scope is less well defined and it would be inappropriate to fix the contract costs. The challenge with managing this form of contract is largely in ensuring that the project scope becomes fixed at some point and the tendency for continual change is avoided.

An appropriate type of contract will have been defined as part of the project procurement strategy and the focus during delivery is on the management part of the strategy (Figure 5-10).

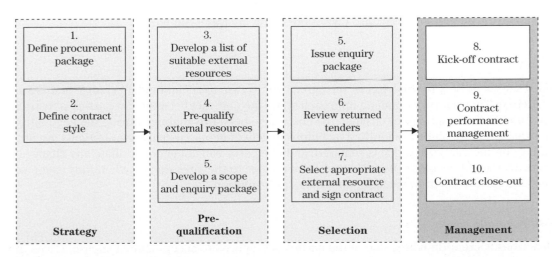

Figure 5-10 Procurement strategy

Once the supplier has been selected, final negotiations completed and the contract signed, the management phase tends to focus on supplier management (behaviours and relationships). Indeed, there is a view that resorting to the contract is tantamount to failure of control in the project – but the contract does help define the boundaries of the relationship.

Managing the contract

As shown in Figure 5-10, there are three phases to contract management:

Kick-off

This sets the scene for the remainder of the project. At this stage, the Project Manager should have a clear idea of the contractual obligations he is entering into, in terms of both the expected supplier performance and the supplier's expectations of him.

Contract performance management

During contract execution, the Project Manager should be tracking performance against plan (deliverables quantity, quality, functionality and timing). Whatever metrics are chosen to track this performance, they should be agreed with the supplier.

Contract close-out

This is an often overlooked phase in the contractual relationship, but a formal close-out of the contract will ensure that no loose ends are left. A contract close-out plan should look at variations, scope, deliverables and feedback.

Managing the relationship

At the heart of successful contract management is the relationship between the parties. Supplier (or contract) relationships tend to fall into one of the following three categories:

Adversarial

The parties stick closely to the letter of the agreement, refuse to accommodate change and are seen as obstructive. This is a very common approach when the contract between the parties is restrictive (critical time limits for example) or when the costs have been driven down during the negotiation phase.

Adversarial relationships are typified by failure to deliver scope (usually requirements are not adequate to define everything that is required, and one party will not accommodate any changes required). The tendency is for the project to drive change to the later stages (when they cost more) and to require the involvement of senior stakeholders in resolution.

Complicit

This describes a situation where one of the parties is prepared to accept almost anything from the other. This could be either for a quiet life (which may be a symptom of poor project management) or because the parties are seen to have too close a working relationship - perhaps through working together over a number of contracts.

Complicit relationships are typified by change. One of the parties is likely to accommodate almost any change required (to keep the customer happy), and this may again result in substantial cost related changes appearing late in the project (when the supplier has realized the impact of their accommodation).

Compromise

Both parties accept that change is going to happen on the project and work together to reach an appropriate level of compromise to deliver the project.

Compromise style projects are typified by mutual agreement (or partnering) and tend to accept the need for change and build that into the project during execution. Compromise style projects are probably the most beneficial to overall project performance, but it needs to be understood that they have a tendency to drift into a complicit style ('…we thought you wanted that…') with attendant risks.

On anything but the smallest project, a number of different relationship styles will exist, and the Project Manager needs to establish an appropriate management style and behaviour for each relationship. Table 5-8 illustrates some of the potential techniques in dealing with the different types of contractual and relationship style.

Table 5-8 Contractual and relationship style

Style	Project management focus	Risks	Mitigation
Adversarial	Scope	Scope inadequately defined. Supplier scope not well understood	Formal review of supplier scope. Highlight gaps and proactively seek confirmation/change
	Change	Multiple changes from supplier post contract award	Formal review of supplier scope. Establish rapid review of change requests and formal feedback to supplier. Aim is to indicate that the detail is understood.
Complicit	Scope	Scope drifts without change requests – may result in multiple changes being revealed late in the project	Ensure formal sessions to refine and detail scope as project is developed (Chapter 4). Use Scope Tracker to challenge variations which do not add value.
	Deliverables	Number of deliverables increases without any consequent change in project benefits or value	
Compromise	Scope	Scope drifts without change requests – may result in multiple changes being revealed late in the project	Ensure formal sessions to refine and detail scope as project is developed. Use Scope Tracker to challenge variations which do not add value.
	Change	Multiple changes raised to accommodate new features into the project	
	Deliverables	Number of deliverables increases without any consequent change in project benefits or value	

Contractual conflict

Irrespective of the management style adopted, at some point during the project, contractual conflict is likely. The Project Manager's best tool in the case of conflict is logic and reason. Sticking rigidly to the line 'the contract clearly states that…' is unlikely to resolve the conflict – even if it really does say that.

Follow a formal process to work with the supplier to define the root cause of the conflict (e.g., the '5-Whys' process). In many cases, the conflict will result from:

- *Poorly defined requirements* – Perhaps it was not clear what was really needed, or the requirements have changed in an uncontrolled manner. Be conscious of the detail of the requirements. The overall requirement may be the same (a new office layout), but the detail (six workstations and a drop-in area instead of eight workstations) may have a real impact on the supplier.
- *Poorly understood requirements* – Did the supplier miss something important? If so, what can be done to address it? Conflict will result if it becomes clear that the supplier will not stand the impact (cost or schedule) on his own. Compromise is the best option – with the prospect of negotiated resolution better than a stand-off.
- *Aggressive contractual bidding/negotiation* – Was the price too low and did the supplier expect to make up for the low bid on variations (which are not happening)? Identify the critical items (particular pieces of equipment for example) within the scope and establish a mechanism to limit the risk to the overall project. An example may be to consider a variation in one critical area, whilst looking for a consequential benefit in another, less critical, area.

Contract relationships need not be all negative. In a bid to overcome the potential issues with traditional forms of contract, many companies now use alliance or partnering agreements. Care need to be taken that this approach is not used as a euphemism for reducing the cost of the contract, without changing behaviours. The general philosophy is that:

- The project objectives, benefits and business case need to be sufficiently transparent for all partners to understand them.
- A single Project Team needs to be formed to combine the best talents from each of the partners.
- There is no need for detailed supervision of supplier team members within the project (so called man-marking).

The Project Manager needs to manage the whole team together to achieve the project aims, and successful partnering agreements can deliver projects more quickly and cost effectively than traditional means.

Contract tracking and supplier performance management

In order to control the overall management of suppliers, two associated processes need to be completed by a Project Manager:

- *Overall management of the contract plan* – progressing contracts in line with the needs of the project: cost, schedule and scope.
- *Individual management of each separate supplier* – developing an appropriate working relationship which enables the supplier to deliver to the project requirements.

Two typical tools are outlined to support these processes:

Contract tracking

Contract tracking is one way of identifying the impact of a particular contract on a project. It focuses on the externally generated project deliverables and their progress. The majority of contract tracking tools provide a method to identify contract risk and therefore mitigate it.

Supplier performance management

Supplier performance management is the process of managing the relationship with a supplier. It focuses on assessing supplier performance through defining performance criteria crucial to the project and then tracking those criteria. The majority of supplier performance management tools provide a method to identify supplier performance risk and therefore mitigate it.

Tool: Contract Tracker

The aim of the tool (Table 5-9) is to rapidly identify if any major contract is out of control and what risk that poses to the project.

Table 5-9 The Contract Tracker explained

Delivery Toolkit – Contract Tracker						
Project:	*<insert project title>*			**Project Manager:**		*<insert name>*
Date:	*<insert date>*			**Page:**		*1 of 1*
Contract	**Type**	**Supplier**	**Contract tracking**			
			Planned status	**Actual status**	**Threat to project**	**Mitigating action**
<insert the procurement package>	*<insert contract style>*	*<insert supplier name>*	*<insert %>*	*<insert %>*	*<insert low, medium, high>*	*<insert any mitigating actions as a result of the status>*
Summary contract status						
<insert summarizing comments on the effectiveness of the contract delivery>						

Contract

Insert the package which is to be procured. Typically, this will only be for major packages, but do not make the package too large and complex (e.g., 'supply of new facility') as the tracking will be meaningless. Try to establish the largest, indivisible packages.

Type

What type of contract is being used (fixed price, reimbursable, etc.)

Supplier

Who is the supplier?

Contract tracking

How is the contract proceeding versus plan at the moment?

- Planned status – How far should contract have progressed at this point?
- Actual status – How far has the contract progressed?
- Threat to project – What are the threats to the project from this contract failing to deliver. If a gap in progress has been identified is this a threat?
- Mitigating action – What actions will be taken to resolve the threat and keep the contract and project on course.

Tool: Supplier Performance Management Tool (parts a and b)

Supplier Performance Management is a 2 part tool. It provides an indication of the performance of a supplier and highlights areas where the Project Manager and team may have to concentrate their control efforts.

The performance of major suppliers within the project is critical to its success. An initial view of supplier performance is often established at the supplier selection stage. This will provide a snapshot of the supplier's likely performance. Subsequently during contract delivery the performance of the supplier can be evaluated against this baseline.

The 2 parts of this tool are:

- Part a – Supplier Performance Evaluation
- Part b – Supplier Performance Tracker

Part a – Supplier Performance Evaluation

This tool (Table 5-10) provides a formal scoring system for each supplier.

Required supply

It is useful to include a short statement about what the supplier is providing under the contract and what impact the supply has on the project.

Performance area

The supplier performance is scored under a number of subheadings which are considered to be general for all projects. Specific performance areas may not be relevant to some projects and may be ignored – the intention is to provide a score relative to the desired (or best in class) performance for each relevant area.

Project management and control

This performance area focuses on whether the supplier has shown adequate project management and control. The evaluation of this area will tend to focus on the interface between the Project Team and the supplier. Typical questions that can supplement the ranking in Table 5-10 are:

- Does the Project Manager appear to know what is happening?
- Do change orders appear without adequate justification or late in the project?

Table 5-10 Supplier Performance Evaluation explained

Delivery Toolkit – Supplier Performance Evaluation				
Project: <insert project title>		**Project Manager:** <insert name>		
Date: <insert date>		**Page:** 1 of 2		
Supplier: <insert supplier name>		**Contract Ref:** <insert reference>		
Required supply				
<enter details of what the supplier is providing under the contract>				
Supplier performance assessment				
Performance area	**Ranking**			
Project management and control	» *Non existent or inadequate project management* » *No evidence of formal controls in place*	» *Project Manager identified* » *Simple project plan in place, no evidence of update/use* » *Reliance on individuals rather than process* » *Haphazard controls*	» *Project Manager and team identified* » *Project plan in place but limited evidence of update/use* » *Project team managed through meetings – no use of tools* » *Inconsistent change control*	» *Project Manager has formal ownership of project and team* » *Single point of accountability for decisions* » *Formal project management processes followed* » *Rigorous change control*	» *Full and accountable project management process* » *Detailed and controlled processes* » *Full visibility of progress, issues and changes*
Score	<enter score 1>	<enter score 2>	<enter score 3>	<enter score 4>	<enter score 5>
Communications	» *Difficult to contact or obtain a response* » *Evidence of poor internal comms* » *Response regularly inadequate*	» *Regular comms but often incomplete* » *Response to queries inconsistent* » *Reactive*	» *Fairly rapid response to queries* » *Generally complete responses, but clarification often required*	» *Effective comms and relationships* » *Generally pro-active and complete responses* » *Little clarification required*	» *Excellent, open relationship* » *Complete response to queries* » *Pro-active and anticipates issues*
Score	<enter score 1>	<enter score 2>	<enter score 3>	<enter score 4>	<enter score 5>
Flexibility	» *Inflexible and reliant on contract*	» *Some willingness to be flexible, but only short-term*	» *Willing to be flexible around project demands over medium term*	» *High degree of flexibility around project and contract matters*	» *Completely open and flexible – joint partnering arrangement focused on project*
Score	<enter score 1>	<enter score 2>	<enter score 3>	<enter score 4>	<enter score 5>

(Continued)

Table 5-10 (Continued)

Delivery Toolkit – Supplier Performance Evaluation				
Project: <insert project title>			**Project Manager:** <insert name>	
Date: <insert date>			**Page:** 2 of 2	
Supplier: <insert supplier name>			**Contract Ref:** <insert reference>	
Supplier performance assessment				
Performance area	**Ranking**			
Capability	➤ Inadequate capability ➤ Consistently missing critical deadlines or milestones ➤ Multiple design or production errors	➤ Poor capability ➤ Some missing of critical deadlines or milestones ➤ Design or production errors not satisfactory	➤ Satisfactory capability ➤ Almost no missing of critical milestones or deadlines ➤ Design or production errors not critical	➤ Good capability ➤ No missing of critical milestones or deadlines ➤ Virtually no design or production errors	➤ Excellent capability ➤ No missing of any project milestones or deadlines ➤ No design or production errors
Score	<enter score 1>	<enter score 2>	<enter score 3>	<enter score 4>	<enter score 5>
Delivery	➤ Frequently capacity constrained resulting in significant schedule problems ➤ Expediting regularly required	➤ Some capacity constraints with some impact on schedule ➤ Some expediting required	➤ Generally unconstrained and able to meet schedule ➤ Limited expediting required	➤ Regular deliveries on schedule ➤ Limited capacity to reschedule to meet project changes ➤ Little or no expediting required	➤ Established track record of deliveries ➤ Capacity to reschedule to meet project changes ➤ No expediting required
Score	<enter score 1>	<enter score 2>	<enter score 3>	<enter score 4>	<enter score 5>

➤ Is the project schedule realistic and achievable? Is the schedule too simple to be meaningful or too complex to be controllable?

It is important in this evaluation process to ensure that the project management and control *process* is being evaluated and not the personalities. Personalities can have a significant impact on the project and need to be dealt with, but may disguise either perfectly good supplier performance (via conflict) or failure (via overfriendliness).

Communications

This performance area extends the project management and control area into general communications. Personalities can have a significant impact on communications, and this ranking exercise can highlight problems that are rooted in an individual. An example would be hoarders of information who become bottlenecks in the communications process.

Flexibility

This performance area looks at how flexible the supplier is to project change. It is important to note that supplier flexibility rarely comes without a price tag. Whilst the Project Manager may desire ultimate flexibility in the supplier, there needs to be an element of reasonableness in the assessment.

What the Project Manager should be looking for is the willingness of the supplier to work with the project, rather than his willingness to implement change for free.

Capability

This performance area looks at one of the key aspects of any successful project: is the supplier delivering a fit for purpose solution? The ranking in this performance area needs to be established on the basis of an objective assessment of, for example:

➤ Number of design errors in documentation submitted.
➤ Number of failures of equipment during performance testing.

Delivery

This performance area is relatively straightforward. Although a supplier may have only one major project deliverable (a piece of equipment or a completed training manual), milestones within the project can be used in the ranking, for example, submission of documents for review or even invoices. Often it is useful to categorize delays as critical or non-critical in terms of their impact on the project.

Ranking

The ranking system uses a scoring from 1 to 5 with 1 as the lowest. An appropriate score should be entered against each performance area.

The ranking process is best carried out with key members of the Project Team who have routine interaction with the supplier. A brainstorming session lasting up to one hour would be sufficient for most suppliers. Assessments must, however, be objective. Scores of 1 or 5 need to be supported by evidence of inadequate or exceptional performance to avoid prejudice generating the results.

Part b – Supplier Performance Tracker

Once the ranking has been completed, the Supplier Performance Tracker (Table 5-11) should be used to:

➤ Plot the evaluation scores on a radar or spider chart (an example is included in Table 5-11).
➤ Compare the supplier performance with a benchmark performance (this has usually been generated as a part of supplier prequalification or selection processes).

This tool gives a visual indication of supplier performance against either the desired or best in class profile.

Supplier profile

The supplier profile generated by the evaluation should be plotted on a radar or spider chart. This approach allows the profile of supplier performance to be ranked against an ideal state and against

Table 5-11 Supplier Performance Tracker explained

Delivery Toolkit – Supplier Performance Tracker			
Project:	*<insert project title>*	**Project Manager:**	*<insert name>*
Date:	*<insert date>*	**Page:**	*1 of 1*
Supplier:	*<insert supplier name>*	**Contract Ref:**	*<insert reference>*

<table>
<tr><td colspan="4" align="center">Supplier profile</td></tr>
</table>

<insert profile generated from the Supplier Performance Evaluation – see example>

<insert agreed action based on the gap analysis – reference versus actual supplier profile>

earlier supplier performance assessments. The tracking chart may then form part of a visual factory showing performance against key procurement packages for the project.

Supplier action plan

The supplier action plan takes the result of the evaluation and details how the project will remedy any gaps between the reference (ideal) and actual performance. Specific actions should be included, detailing timescales and responsibilities. Finally, a date for review of the supplier post-action plan should be scheduled.

Use of the Supplier Performance Management Tool

This tool is most appropriate in two situations: when the project (and supplier involvement) is of sufficient duration long for a reasonable history of performance to have built up; and as a post-project review tool.

When used within the project, the tool is at its most useful when applied at specific milestones. A likely scenario for a typical equipment supply project would be:

- *Order placement* – Benchmark is probably based on experience of the supplier, any audit results and the current project procurement process (how and why the supplier was selected this time).
- *Design completion* – This would usually follow a sufficient period of time for a relationship to have been established during the kick-off and early phases of contract execution.

⫸ *Equipment testing* – Again, sufficient history of the contract execution will have been gained to understand any performance issues.

At each milestone, the Project Team would use the ranking to establish if any risks to the project exist and what mitigating actions would be required. For example, if poor communications were identified at the design stage, the Project Manager may decide that temporary secondment of a Project Team member to the supplier's office would be needed to mitigate this risk in future project stages.

Short case study

Company A engaged a respected automation supplier to design, build and install a replacement control system for an aging plant area. The company had used this supplier in the past on a number of its sites, and the supplier performance had been technically excellent, but there were some concerns over the project controls.

At contract kick-off, the Project Manager emphasized the need for clear controls, and the supplier indicated his understanding and explained the control mechanisms which would be applied. This was deemed satisfactory and the contract proceeded without any problems being anticipated.

The contract seemed to go well initially, with few issues escalating to the Project Manager's attention. The design process for this contract was phased, and there was no single 'design complete' stage. The Project Manager, therefore, decided to carry out a supplier performance evaluation part-way through the design. The evaluation was carried out with the whole Project Team, and the results are shown in Table 5-12.

Table 5-12 Case study supplier performance evaluation

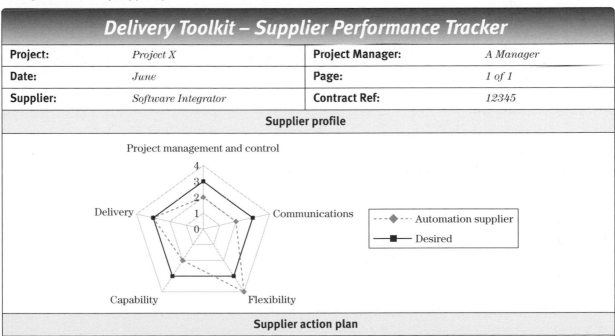

Delivery Toolkit – Supplier Performance Tracker			
Project:	Project X	**Project Manager:**	A Manager
Date:	June	**Page:**	1 of 1
Supplier:	Software Integrator	**Contract Ref:**	12345

Supplier profile

Supplier action plan

Supplier team appears to show no capability in delivering to our requirements and is not controlling or communicating clearly. However, the supplier is flexible to our requirements. A formal project review is required to resolve the issues so that specific actions can be put in place and tracked.

This evaluation (involving the whole team) brought out the first indicators of problems in three areas of the project: management and control, communications and capability. In addition, it showed that the supplier was highly flexible to project demands.

In each of these areas, it was clear that the Project Manager was unaware of the problems that existed at the technical level where there was a mismatch between the supplier's designs and the company engineers' expectations.

As a consequence of this evaluation, a project review meeting was requested with the supplier. It emerged that although he had been highly flexible, he was working without effective controls. The supplier was, in fact, now forecasting an increase in project spend of approximately £120,000 through work completed without any change authorization (representing an approximate 10% cost increase).

As a result of the review, the supplier changed his Project Manager and put more effective controls in place to ensure that change was adequately controlled. In addition, this new supplier Project Manager regularly reported to the company Project Manager as well as developing a strong working relationship with the company technical lead on this package. Technical mis-communication was solved by a closer working relationship but within a change control framework.

Project control strategy

In the project planning phase, the cost schedule and change plan will have been developed against the project requirements. These plans will define the control strategy as shown in Figure 5-11.

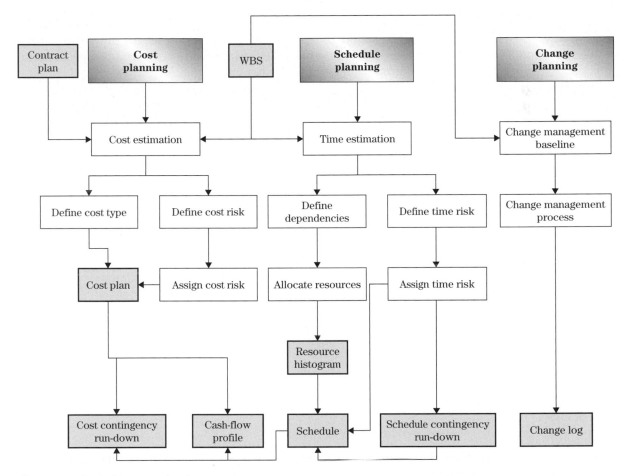

Figure 5-11 Control strategy development process

Project control involves taking this strategy and the associated tools and pulling them together to maintain control.

The most visible parts of the Project Manager's activities are the monitoring and reporting of cost schedule, change and, most importantly of all, forecasting the outcome. Cost plans, cost reports and schedule updates show a historical view of the project. The project management skill is in taking these reports and using them to forecast the outcome of the project with a high degree of certainty. (When will it be finished? What will it cost? What will I get?)

Forecasting is notoriously difficult if the basis of the forecast is flawed. Areas that the Project Manager needs to be concerned with are:

- *Lags in the flow of information* – For example, how will man-hours be recorded and costed to the project or when will invoices be charged? In many cases, the formal cost reports for a project can be at least 1 month out of date. The Project Manager needs to understand the historical reporting basis for the costs, so that the forecasts make sense.
- *Schedule reality* – If a project schedule never changes during the life of a project, then it is unlikely that the project will complete on time. The project schedule should not be a static representation of what was intended, but a dynamic picture of reality. Many tools exist to review the performance against schedule as detailed in *Real Project Planning* (Melton, 2008). The important control feature is that it accurately forecasts the project future and its outcome.
- *Allowing change* – Not all change is bad. A project will be subject to periodic changes in, for example, scope, deliverables, time or cost. Being in control does not mean preventing change; it does mean controlling change appropriately and understanding the impact of change on the outcome.

Defining control

Leading indicators

Keeping control does not mean simply tracking what has happened. It also means being able to predict whether you will remain in control and predict the final project outcome.

Within any project, it is important to identify key measures which can be used to forecast whether you are in control or not. Some of these measures, for example Earned Value Analysis (EVA), help define whether the project will achieve its objectives within schedule and cost objectives. Other measures are more useful in indicating whether the quality aspects of the project will be met.

Typical measures which can be used in this context are:

- Number of document revisions before final approval.
- Number of changes issued.
- Number of test failures/rework.
- Staff turnover.
- Risk exposure.
- Resource spend rate (cash or man-hours for example).

Used in the right way, each of these indicators helps make a prediction of how well the project is being controlled. Taking each in turn:

Number of document revisions before final approval

If a document (For example, a design specification, test protocol or standard operating procedure) requires repeated revision before it is deemed acceptable, then this is a indicator of one or all of the following:

- *Inadequate initial requirements* – The requirements are not defined well enough to allow the specification to state what is required in a clear manner. For example, a user requirement for a new piece of software may require multiple iterations of the design specification to elicit the requirements adequately.

- *Supplier under-performance* – The requirements are adequate, but the supplier is incapable of correctly interpreting those requirements and repeatedly issues inadequate documentation.
- *Unrealistic requirements* – The requirements cannot be met within the limits of the project scope, and the iterations of the design are indicative of the attempt to reach a compromise.

The risks highlighted by this indicator are that assumptions about the scope are incorrect and change will be required. Because of these various potential issues, it is essential that a root cause assessment be undertaken (possibly using the 5-Whys technique) so that the right problem is addressed.

Number of changes issued

Project change is another leading indicator of control. A large number of changes indicate a potential issue with either the scope of the project (inadequately defined) or the performance of the supplier (not in control) or the performance of the team (scope creep).

Number of test failures/rework

When a system or package within a project is subject to repeated testing or rework, then there is ultimately a control problem. Again the root cause needs to be assessed carefully as it could have been caused by:

- *Supplier under-performance* – The realization of the requirement is poor.
- *Equipment inadequacy* – The requirement cannot be realized by the specified equipment.
- *Requirement interpretation* – The supplier produced what was asked for, but the requirement had been poorly communicated. This is typical of a situation when an end-user representative joins the project at a later stage and applies his interpretation to the requirements, '…That's not what I want…'

Staff turnover

Particularly on large projects, staff will leave and join at regular intervals. Excessive staff turnover is, however, a leading indicator of a number of potential problems:

- *Conflict* – The team is not working properly, either through individual conflict or the Project Manager's approach. In this situation people tend to leave the project by choice.
- *Capability* – The team cannot deliver the requirements, perhaps because the skill mix or the individual competence is inadequate. In this situation people tend to be asked to leave the project.

Risk exposure

Are project risks increasing or decreasing over time? A project in control will tend to decrease the number of high or medium risks as the project proceeds.

Resource spend rate

What is the rate at which man-hours or cash is being expended? Is the spend rate very high compared to schedule progress? A project in control will tend to have a smooth rate of planned spend as the project proceeds.

Lean measures and the use of leading indicators

Once the leading indicators have been identified. It is necessary to choose an appropriate way of using those indicators to highlight problems and drive improvement in the project.

One such method is Six Sigma, which has traditionally been used in a manufacturing environment. The use of Six Sigma processes is becoming common in the service sector, and it can be readily used in the monitoring and improvement of project management processes.

Six Sigma techniques provide a mechanism for using the data collected from processes to demonstrate what is normal for the process and predict what is likely to happen in the future. The systems rely on the use of statistical probability to determine with a level of confidence whether a process is in control, with any fluctuations due to normal operation, or out of (or heading out of) control, indicating that something has changed within the process, with either good or bad consequences. What is normal operation for the process may be inside or outside the targets that the business requires. For example, in tracking the number of revisions required for documents produced by the mechanical design team of an engineering project, what is normal for this process can be determined, thus allowing a target performance level to be set. The process can then be monitored to look for indicators that performance remains at the required level.

Six Sigma measures are based around the concept of a 'process' with defined inputs and outputs. This concept is applicable to most types of process if the process, opportunity and measures are correctly defined. Within projects, the process is the project roadmap which can be generalized:

➤ *Scope definition* – What do we want the project to deliver?
➤ *Design* – How will the project deliver?
➤ *Implementation* – Delivery of the project.
➤ *Testing and handover* – Delivery of the benefit.

At each point within this process, a transformation takes place (ideas to scope, scope to design, design to deliverable and deliverable to business use). The opportunity at each point is therefore the deliverable (documents or package, for example), and the measure is defined in the appropriate leading indicator. Table 5-13 shows how the process and leading indicators relate.

Table 5-13 Process, leading indicators and measures

Process	Leading indicator	Measure
Scope definition	Document revisions	Number of versions before document approval
Design	Document revisions	Number of versions before document approval
	Changes	Number of changes per package or system over the duration of the design phase
Implementation	Document revisions	Number of versions post approved document
	Changes	Number of changes per package or system over the duration of the implementation phase
Testing and Handover	Test failures	Number of tests which failed or required rework as a percentage of total test conducted
	Changes	Number of changes per package or system over the duration of the test and handover phase
Overall	Risk exposure	Number of high risks over the duration of the project
	Staff turnover	Sum of number of staff leaving and joining over project duration
	Resource spend rate	Weekly rate at which man-hours, cash or similar resource is being expended

Not all these measures will be appropriate for all projects, and the Project Manager should pick those considered appropriate for the risks of the project. For example, a project with novel technology may focus on change and document versions, whilst one dealing with a new supplier may focus on document versions and test failures.

Once the measures are established, the Project Manager can make use of Six Sigma techniques to track and manage those measures through the use of many tools. Three examples are covered here:

- Defects per million opportunities.
- Histograms.
- Control charts.

Defects per million opportunities (DPMO)

Defects per million opportunities (DPMO) is a universally accepted way of classifying numbers of defects and, in a project management context, can be used:

- To compare one project with another.
- To compare similar types of project processes.
- As a tool for measuring improvement following a change.

The first step in this process is to define what a defect is and determine what an opportunity is. Table 5-14 provides examples of how an organization could define the criteria and set limits.

Table 5-14 Examples of defects and opportunities definition

Process	Leading indicator	Definition of a defect	Definition of an opportunity
Scope definition	Document revisions	More than 5 revisions before the scope package is approved	Number of scope documents
Design	Document revisions	More than 3 revisions of any single document or drawing	Total number of drawings and documents
	Changes	More than 2 changes per package or system over the duration of the design phase	Total number of packages or systems
Implementation	Document revisions	More than 2 versions of any single post approved document	Total number of documents or drawings
	Changes	Requirement for any changes per package or system over the duration of the implementation phase	Total number of packages or systems
Testing and Handover	Test failures	More than 2 tests which failed or required rework per test conducted	Total number of tests
	Changes	Requirement for any changes per package or system over the duration of the test and handover phase	Total number of packages or systems
Overall	Risk exposure	More than 3 high risks over the duration of the project	Total number of identified risks
	Staff turnover	More than 2 members of staff leaving and joining per month over the project duration	Total headcount required for the Project Team
	Resource spend rate	Actual spend rate differs from forecasted spend rate by more than $10k in any given month	Number of months for the project

This information can be used in several ways once a target DPMO has been set for a particular project or process:

➤ *Project performance during delivery* – DPMO can be calculated at any time during the project lifecycle to see how it is achieving against targets (allowing appropriate action to be taken where necessary).

➤ *Project performance at completion* – DPMO can be calculated for a particular project or process. Historical data can be used to calculate the typical DPMO for all the projects that the organization has completed. Then the two are compared.

Histograms

Histograms are used to look at trends within numerical data. The data points can either be plotted individually or placed into groups. An important consideration when preparing a histogram is to ensure that the resulting bar chart does not contain so much information that trends in the data cannot be identified (there is simply too much information to see what is going on), and also that the sizes of groups is not so large as to bury an issue. Leading indicators can be plotted in this way to show issues within the project. For example, in a business change project where one of the deliverables is a revised set of standard operating procedures (SOP), the number of SOP versions per approved document can be easily collected and plotted as a histogram (see Figure 5-12) to show the overall performance of the project. In this example the revision history for the 18 SOPs is plotted.

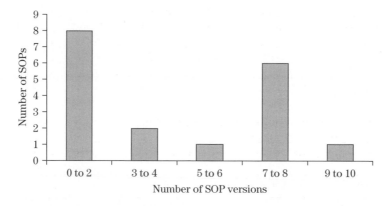

Figure 5-12 Example SOP approval histogram

Whilst histograms are very simple tools, they can be extremely powerful in taking lists of numbers with no discernable pattern and producing a graphical representation allowing patterns and trends to be identified. From Figure 5-12, the following conclusions can be drawn about the data. Approximately one-third of documents are completed with only one or two revisions.

➤ There are relatively few which require three to six revisions
➤ A large number require seven or eight revisions.
➤ 10 SOPs are within the company average (4 versions or less) a significant number are higher.

Looking at the data in this graphical representation provides a starting point for a more focused study of the business process. Examples of the type of conclusion that can be drawn from this, depending on the actual scenario, could be:

- The organization has a good understanding of some areas of the project but a very poor understanding of others. Documents for well-understood areas require very little revision, whereas those for the poorly understood areas require many revisions. The design process for these areas could be stopped temporarily whilst the root cause issues are defined and dealt with.
- A single or small group of individuals may have a very poor route of communication to the project experts and therefore produce standard documents which are not fully appropriate for the specific project.

The information from the bar chart can also be used to draw other conclusions about the project process. For example, if earned value is being used to review man-hour usage against progress, it can start to identify where potential problems will occur if targets are not being achieved.

Control charts

Control charts are used to review the performance of a process over time. They identify whether a process is in control and capable, whether the process is operating as normal, or whether things have changed which are about to affect performance. Control charts require the use of statistical tools. Many larger organizations will have software to produce the charts; however, the most common spreadsheet packages have the capabilities to produce the necessary information such as mean and standard deviation.

- *Control* – A process is defined as in control when its performance data forms a predictable distribution within control limits. A process is out of control when its data either has distinct rises or falls, is consistently high or low, or not properly distributed around the mean.
- *Capability* – A process which is capable is defined as one in which the specified performance limits are outside the control limits. This means that if the process is in control, it will always meet its specification requirements.

Although control charts are often used in the evaluation of manufacturing processes, they are also applicable to evaluating the performance of the project process. The example in Figure 5-13 shows a particular project process (document generation) in control but which is not capable.

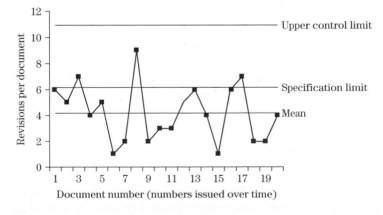

Figure 5-13 Example 'in control' control chart

Figure 5-14 shows what happens as the project progresses from document numbers 20 to 40. The process is not capable and is out of control in two areas:

➤ The seven points after the first point (document number 21) are all less than the mean. Whilst this is probably a desirable effect as it means documents are receiving approval with few revisions, a run of seven points like this is statistically unlikely, and this may indicate a problem which requires investigation. For instance, the team of reviewers and approvers may be busy with other work and not giving documents the attention and consideration they require. This will give rise to issues later when problems which could be identified now have to be resolved during construction or testing.

➤ The seven points starting with document number 30 are all rising. This indicates that something has changed somewhere within the project process and is having a negative effect on its performance. As can be seen in the graph, although the next point starts to fall, the process quickly goes out of its control limit.

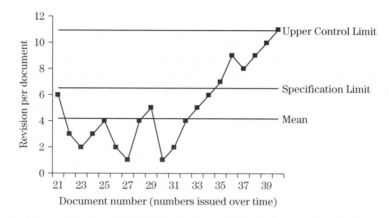

Figure 5-14 Example 'out of control' control chart

As can be seen from the examples, control charting is a powerful and complex tool, which although relatively simple to prepare, can with the right resources, indicate a vast amount about the project processes.

The purpose of illustrating control charting here is to show how it can be used in a project scenario, and only the relevant basic concepts are discussed. For the purposes of preparation and use of control charting, further reading and experience are required (Kiemele et al., 2000).

Control charting is not intended to be an approach to all projects. There is cost/time overhead in setting up the techniques, and on simple projects direct measures are more appropriate. However, when a project is complex or of a significant duration and has a substantial number of packages or deliverables, control charting may provide a means to forecast project performance and establish corrective action.

Tool: Project Control Tracker

Bringing these various control elements together into a single view is the function of the Project Control Tracker (Table 5-15). The aim of this tool is to succinctly present project performance data so that the appropriate response can be made to ensure the project remains in control and able to deliver a successful outcome.

Table 5-15 Project Control Tracker explained

Delivery Toolkit – Project Control Tracker					
Project: <insert project title>			**Project Manager:** <insert name>		
Date: <insert date>			**Page:** 1 of 1		
Cost					
Area	**Control tool**	**Current status**	**RAG rating**	**Action/mitigation and forecast status**	**RAG rating**
<insert area, e.g. Spend as planned>	<insert control tool, e.g. Cost plan>	<insert current status of project cost plan>	<current RAG level>	<insert the action to be applied>	<future RAG level>
<insert area, e.g. Spend un planned>	<insert control tool, e.g. Cash-flow profile>	<insert current status of project cash flow>	<current RAG level>	<insert the action to be applied>	<future RAG level>
<insert area, e.g. Spend risk>	<insert control tool, e.g. Cost run-down or contingency spend>	<insert current status of project contingency>	<current RAG level>	<insert the action to be applied>	<future RAG level>
Schedule					
Area	**Control tool**	**Current status**	**RAG rating**	**Action/mitigation and forecast status**	**RAG rating**
<insert area, e.g. Done when planned>	<insert control tool, e.g. earned value analysis>	<insert current status of project schedule>	<current RAG level>	<insert the action to be applied>	<future RAG level>
<insert area, e.g. Done by who planned>	<insert control tool, e.g. Resource plan>	<insert current status of project resource plan>	<current RAG level>	<insert the action to be applied>	<future RAG level>
<insert area, e.g. Schedule risk>	<insert control tool, e.g. Schedule buffer or contingency run-down>	<insert current status of project buffer>	<current RAG level>	<insert the action to be applied>	<future RAG level>
Change					
Area	**Control tool**	**Current status**	**RAG rating**	**Action/mitigation and forecast status**	**RAG rating**
<insert area, e.g. Staff turnover>	<insert control tool, e.g. Staff turnover control chart>	<insert current status of staff turnover>	<current RAG level>	<insert the action to be applied>	<future RAG level>
<insert area, e.g. Document revisions>	<insert control tool, e.g. Document revision control chart>	<insert current trend of document revisions>	<current RAG level>	<insert the action to be applied>	<future RAG level>
<insert area, e.g.Control change>	<insert control tool, e.g. Change log>	<insert current status of project changes>	<current RAG level>	<insert the action to be applied>	<future RAG level>
Project control summary					
<insert summary comments on the status of the project and impact on the outcome>					

Within the cost, schedule and change sections the following guidance applies:

Area

This refers to the control area for the project (Figure 5-11):

- *Spend as planned* – Is the project spending in line with plans, or too much, or too little? (The latter two indicate a potential loss of control.)
- *Spend unplanned* – What unplanned spending has the project incurred?
- *Spend risk* – What is the risk of the project going over budget?
- *Done when planned* – Are tasks completed when planned or delayed?
- *Done by the person who planned* – is the resource available to complete the project?
- *Schedule risk* – What is the risk of the project going over schedule?
- *Control change* – How are changes being captured and controlled?

Control tool

Insert the tool that will be used to monitor whether the project is in control or not. The examples given in the table are typical of those used on many projects:

- *Cost plan* – What have I planned to spend and am I in line with the planned spend?
- *Cash flow profile* – What have I actually spent against planned cash flow?
- *Cost run-down (or contingency spend)* – How much contingency have I spent to date?
- *Earned Value Analysis* – What is the value of the work actually done?
- *Resource plan* – Who is working on the project compared to the plan?
- *Schedule buffer or contingency run-down* – How much buffer do I have in the schedule?
- *Change log* – How many changes are in the project? How many approved or rejected?

Current status

Explain the current status of each of the control areas? Insert a few words to summarize the position and record using a RAG status for ease of reference. The RAG levels should indicate:

- *Green* – The project is in control.
- *Amber* – There are some problem areas, but mitigation steps are in place and the project remains in control.
- *Red* – Without action, this area of the project will be out of control and therefore potentially impact project success.

Action/mitigation and forecast status

What actions will be taken to restore or maintain control? The forecast status should indicate what the impact of the action will be on the current status.

Short case study

A project was underway for the transfer of a bulk pharmaceutical product from one site to another to release production capacity at the original site. The project had a significant initial budget and a

Table 5-16 shows an example of using a project control tracker.

Table 5-16 Example Project Control Tracker

Delivery Toolkit – Project Control Tracker					
Project:	Confidential		**Project Manager:**	Bob Smith	
Date:	June		**Page:**	1 of 1	
Cost					
Area	**Control tool**	**Current status**		**Action/mitigation and forecast status**	
Spend as planned	Cost plan	Current spend is ahead of plan and indicates that the project cannot be delivered for the original budget	R	Budget needs to be revised and a new funding request raised	A
Spend un planned	Cash flow profile	Additional spending on contract engineering resources to meet timelines	A	Budget needs to be revised and a new funding request raised	G
Spend risk	Cost run-down (contingency spend)	Contingency spend will be required to maintain progress without additional funds	A	Budget needs to be revised and a new funding request raised	G
Schedule					
Area	**Control tool**	**Current status**		**Action/mitigation and forecast status**	
Done when planned	EVA	Project on status	G	None	G
Done by who planned	Resource plan	Issues with adequate skilled resource – having to source additional personnel	A	Revised resource plan	G
Schedule risk	Schedule buffer or contingency run-down	Project buffer remains unused	G	None	G
Change					
Area	**Control tool**	**Current status**		**Action/mitigation and forecast status**	
Staff turnover	Staff turnover control chart	No staff left or joined team	G	None	G
Document revisions	Document revision control chart	Document revisions showing a trend towards more revisions than company benchmark	A	Identify cause of trend and establish correction mechanisms	G
Control change	Change log	Existing plant utilities not capable of supporting requirements – significant change to scope	R	Revised plan and budget to accommodate change	G
Project control summary					
Project remains on schedule, but during design the utilities plant at the receiving site needs to be upgraded to accommodate product. This will require additional funds					

very tight timescale, as the original site was capacity constrained and could not supply the market adequately without this product relocation.

Taking schedule as the key driver for the project, the Project Manager allocated resources to drive the project forwards. Table 5-16 shows the status of the Control Tracker shortly after a major design issue was discovered and analyzed. During the design review the utilities plant at the receiving site was discovered to be incapable of handling the transfer product requirements and the cost plan had to be reviewed to cope with this major scope addition.

The project appeared to be on schedule, but a clear change had been identified which required additional funding to meet the overall project needs. These additional funds required the Project Team to revisit the business case and confirm that the project benefits were still applicable with the additional costs. Having a balanced control strategy supported the Project Manager and Project Team in highlighting the full impact of the change and the fact that they were in control – the Control Tracker had helped to identify and communicate the issue.

Project reporting

Reporting on project status is closely allied to the control strategy. The Control Tracker forms a useful vehicle to generate project reports as it can provide the full breadth of control data needed for all audiences (Figure 5-15). Lean reporting is the process of using the same base data for different audiences to generate the required action/reaction that supports project progress.

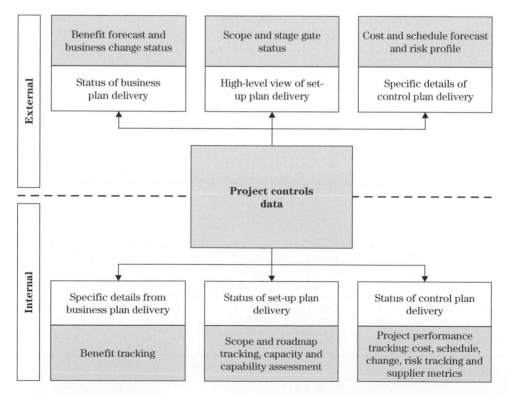

Figure 5-15 Lean reporting

Historically, project progress reports have been long on detail and (sometimes) short on clarity. It is easy to generate pages of information which try to convey the status of the project but end up losing the main message.

A lean report does not necessarily mean a short report. Lean means removing everything that does not add value to the final product (in this case, to convey meaningful information about the current and future state of the project). Lean reporting is about extracting the key pieces of information appropriate to the intended audience. As summarized in Figure 5-15, an internal audience (the Project Team) has quite different control data needs from an external audience (the sponsor and the business).

Whoever the audience is, a lean report should contain some consistent information on:

- What is the project's overall goal? – Why are we doing this project?
- What is the status of CSFs? – Failure to achieve these will impact the goal. This is the point at which risk can be highlighted.
- How is the project progressing against plan and what is critical to progress? – Perhaps the number of contracts placed against the plan or the earned value of activities.
- How is spending progressing against plan? – Is cash flow in line with expectations? Has the contingency been used?
- How controlled is the scope of the project? – How many changes have there been?
- How is the business responding to the changes the project is delivering? Perhaps in terms of training or engagement with business process change activities?

A simple report illustrating this approach is shown in the case study in Chapter 7. Detailed information on the status of individual deliverables (or detailed resource plans) are more appropriate as control views on the data within the Project Team, rather than a report to others outside the project. For internal team reporting, visual techniques can be used rather than traditional documents. For example:

- *Project intranet* – an e-environment containing all the detailed progress information that team members need to access to perform their role.
- *Project notice boards* – using traffic light indicators to highlight specific types of action to specific team members. For example, the project critical milestone schedule (those milestones on the critical path) may be highlighted as red, amber or green depending on the status. Team members know that if a milestone they 'own' changes to amber, then they have to commence mitigation plans.

The way performance is measured and reported to various audiences will impact the things that people do and the way they behave. This will contribute to the project culture and the culture of project management in the business:

- Culture = how we behave + values we hold + things we do (Figure 5-16).
- To change culture we need to change these.

In a project context, there is a need for reports and method of reporting that will drive the right behaviours, reinforce the right values and, therefore, to ensure that actions are in the best interest of the project and the business.

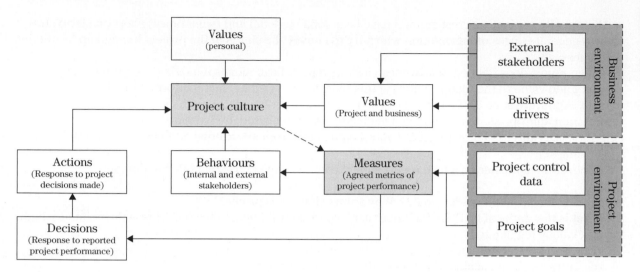

Figure 5-16 Project culture

Project review

Project reviews are an important part of the learning process and are appropriate at various points in the project, not just at the end. The three types of project review that are typically undertaken are covered in this section:

- *Project delivery reviews* – These are the usual project performance reviews conducted as the project is being delivered which support assessment of current performance and advise change in order to improve that performance as necessary.
- *Project completion reviews* – These are the typical end-of-project review sessions to establish overall project performance (historically) and generate and share knowledge.
- *Project crisis reviews* – These are reviews conducted when a major issue occurs, either during a project or when a delivered project is deemed to have failed and requires further analysis.

The same themes are seen in all types of project reviews:

- *Generate knowledge* – What works and what does not.
- *Share knowledge* – Repeat good things and do not do the bad things again.
- *Modify future practices* – respond to issues in a timely manner, either within a project or within the wider organization.

The intention of all project review methodologies and tools is to understand a situation and learn from experience so that the next project will not repeat the mistakes of the last. Ideally an organization will gain sustainable knowledge so that the next project repeats all the good practices and not of the poor ones.

Project delivery reviews

The frequency of project delivery reviews depends to a large extent on the type, length and complexity of the project. At the very least, a formal review should be undertaken at major project decision points or stage gates. For example, during a business change project, it would be usual to conduct a review of the design phase before implementation commenced. This review is separate from the design approval stage gate but allows reflection of the project management performance during design.

Month-end review

Most Project Managers use the month-end reporting process as a method of reflecting on performance as well as to forecast likely project outcome. The exact format and content of a monthly report should have been decided in the planning phase but it would be usual to reflect on:

- *Overall project performance* – Using data generated from the delivery of control strategy. For example, risk rating, progress against schedule and budget, safety and quality metrics.
- *Team performance* – The tangible and less tangible metrics that indicate whether the team is working well or not. Typically, team performance is linked to project metrics on the basis that measures determine behaviours. For example, if a team is praised for always delivering on time with no reward for quality, then they are likely to prioritize schedule over quality when a conflict occurs.
- *Individual performance* – The knowledge, skills and behaviours that individuals display as they contribute to the team effort of project delivery.

The month-end review is a timely way for a Project Manager to modify or change project conditions in order to mitigate a negative trend in project performance before any crisis is reached. For example:

- Individual or team capability development or in extreme cases, changes in individual assignments within or outside of the team.
- Modifying project operating procedures or implementing new ones.

'In control' review

The in-control review is an extension of the Project Control Tracker concept and is a periodic activity designed to confirm that all aspects of the project are in control for delivery. The 'In Control?' Checklist (Appendix 9.3) is covered in detail in *Project Management Toolkit* (Melton, 2007) and provides a mechanism to review the following areas:

- *Stage Two check: Any changes since approval?* This reflects back to the approved PDP (and within that the approved business case) and challenges the review to check that the project aligns with the business.
- *Scope definition: Is it clear, has it changed?* This checks the management of scope change and delivery of scope value. The challenge is to understand the scope delivery decisions that have been made and check their link to the benefits case.
- *Business change management: Is the business keeping its side of the bargain?* This checks that the project has a robust and effective link to the business, as evidenced by activities occurring in the business to 'receive' and 'use' the project outcome. This challenges the effectiveness of stakeholder management.
- *Risk and issue management: Where are the uncertainties in the project? Are plans in place to manage them?* Any review has to check the management of uncertainty and the level of risk which the project is currently operating under, challenging whether the level of mitigation under way matches the level of risk.
- *Project organization: Are the right people in the project and are they working on the right activities?* This reflects on the performance of the Project Manager and the Project Team, the way they are working and the overall morale in the project. The atmosphere in the Project Team meetings or project office can tell a reviewer a lot about the state of the project.
- *Contract and supplier management: What external suppliers are being used? Are they performing?* This reflects on supplier performance and the processes in place to control it. A good reviewer will look out for the signs of adversarial relationships or poor management of challenging suppliers.
- *Project controls strategy: Are the project cost, schedule and changes understood and in control?* This looks at the traditional and less-traditional/lean project metrics. There is still a place for cost and schedule adherence metrics as well as the flow of decisions and approvals, for example.
- *Project review strategy: Are regular reviews conducted and reports generated?* This reflects on how project performance is being managed and whether objective reviews are taking place regularly.
- *Stage Three decision: Is the project in control?* At the end, the reviewer is asked for an objective assessment of the health of the project.

Typically this type of review is conducted by the Project Manager or an independent reviewer (such as a colleague Project Manager) at specific project stage gates. It aims to assess the health of a project at a point in time and can be used to change the management of future project stages so that any risk of failure is mitigated.

Project completion reviews

At the end of a project, there are many methods to conduct performance reviews. Some are more passive than others (for example the traditional close-out report) but still have their place in organizational knowledge management.

Project close-out review

At the end of a project, a Project Manager will traditionally pull together a final project report as a close-out statement. This is as much about project governance as it is about knowledge management. This document, whether one or many pages, tends to be a collation of facts and figures and provides one facet of overall project review. For example, how well specific project metrics were achieved, such as cost or schedule adherence, safety or quality metrics.

After Action Review (AAR)

The After Action Review (AAR) is essentially a team look back over the project in five separate areas:

- *Objectives* – Reflect back on the original CSFs and project objectives. Were they met, did they change and if so, why?
- *Journey* – This is a review of the 'highs' and 'lows' in the project. What were the project high points (team motivation or finding an original solution to the problem) or the low points (when a mistake was made, when a blocker appeared)?
- *What went well?* – Select the things that worked well on the project, and understand why, so they can be translated into learning for the future. A 5-Whys analysis can help get to the root cause and it is this good practice that you want to repeat.
- *What could have gone better?* – For those things that did not work so well on the project, identifying the root cause rather than the symptom is most important. Again it is the root cause which is the poor practice and one which you do not want to repeat.
- *Summary* – Overall, how would the project be summarized to stakeholders or sponsors?

One of the most important parts of this type of after-action review is to ensure that a fair and realistic appraisal of the project is carried out and that individuals are not unfairly blamed for problems. Remember, even if an individual had a major impact on a project, there is usually some systematic cause. It is a good idea to have an independent facilitator for the review to help keep the session on track.

The Project Assessment Tool covered in *Project Management Toolkit* (Melton, 2007) explains the content and aims of the After Action Review (AAR) in more detail.

One aspect of an AAR is increasingly being used to assess the level of waste in the project management process used to deliver the project. This is usually picked up in the *What went well?* (no waste) and the *What could've gone better?* (lots of waste) phases of the review process. The Project Waste Analysis tool (Table 5-17) is one way to systematically review all types of waste.

Tool: Project Waste Analysis Tool

The aim of this tool (Table 5-17) is to support the after-action review process by providing a checklist of typical project waste areas. The key goal for this type of analysis is to identify good and poor project processes and behaviours so that future projects can incorporate this knowledge.

Table 5-17 The Project Waste Analysis Tool explained

Delivery Toolkit – Project Waste Analysis Tool			
Project:	*<insert project title>*	**Sponsor:**	*<insert name>*
Date:	*<insert date>*	**Project Manager:**	*<insert name>*
Team checklist			
Observed waste	**Response**	**Comment**	
1. Over production	*<yes or no>*	*<insert response>*	
2. Waiting	*<yes or no>*	*<insert response>*	
3. Transport	*<yes or no>*	*<insert response>*	
4. Inventory	*<yes or no>*	*<insert response>*	
5. Over processing	*<yes or no>*	*<insert response>*	
6. Motion	*<yes or no>*	*<insert response>*	
7. Defects	*<yes or no>*	*<insert response>*	
Action plan			
<insert any actions to be progressed so that the organization learns from this project>			

Observed waste

Table 5-18 gives examples of the types of project waste seen against the seven standard categories. It also describes the impact of each waste in a project context.

Table 5-18 Project waste approach – people and processes

Waste area	Project impact	Example
Over-production	Project Team members called on to do increasing amounts of work without adding to benefits	Project reporting process requires excessive input from team members which is a distraction from the tasks they would normally undertake
Waiting	People waiting for tasks to be fully completed by others, no clear process to pass on responsibility	There are a limited number of capable resources on the project to tackle specific tasks, causing a bottleneck in the progress of activities
Transport	Process requires that project deliverables are moved excessive distances in order to be worked on	Project team members are not co-located, meaning that the project deliverables (e.g. drawings) need to be shipped between offices and there are no electronic systems in use to mitigate this

(Continued)

Table 5-18 (Continued)

Waste area	Project impact	Example
Inventory	Project processes require a large number of deliverables with limited value	Project required a large number of documents to define the requirements. Many of these documents re-iterated standards and could have been condensed into fewer, specific items
Over-processing	Multiple people required to review and approve deliverables. Limited authority in specific team members	A project supplier required his documents to be reviewed by operations, projects, quality and procurement without any clear single point of authority. The result was that multiple sets of contradictory changes were requested.
Motion	Project Team members required to travel to multiple locations to progress the project, resulting in inefficient performance	A supplier has three offices involved in the development of a new software product and the project requires the team specialist to repeatedly visit the three offices
Defects	Project Team members make undetected errors in system requirements documents	A more junior Project Team member is given the responsibility of defining specific system requirements. Project processes fail to require the documents to be reviewed and approved before issue.

Response

Was waste observed? It is likely that some projects have seen many of these different categories of waste. Try and focus on those which exhibit repeating issues to aid learning.

Comment

Define where the waste was observed and root cause the reason, so that there can be learning development for the organization.

Short case study

A project for the implementation of a new control software system has recently been completed, and a formal review of the project was undertaken using the Project Waste Analysis Tool (Table 5-19).

Project crisis reviews

A project can be 'in crisis' for any number of reasons and at any stage in its lifecycle. There are a number of definitions which apply. A project crisis can be an unstable situation of extreme difficulty; project chaos or a crucial stage or turning point in the project.

It can equally refer to a past event where the crisis was not initially detected. Therefore, crisis reviews during and post-delivery are to be considered.

Table 5-19 Example Project Waste Analysis Tool

Delivery Toolkit – Project Waste Analysis Tool			
Project:	Control System	**Sponsor:**	Sponsor
Date:	January	**Project Manager:**	Project Manager
Team checklist			
Observed waste?	**Response**	**Comment**	
1. Over production	No	N/A	
2. Waiting	Yes	Project expertise in the technology to be employed was limited, so the Project Team were delayed by lack of resource	
3. Transport	No	N/A	
4. Inventory	Yes	Project quality representative required multiple documents structured around the needs of the quality organization. This meant that documents which would have been combined were split into multiple issues, each requiring document control.	
5. Over processing	No	N/A	
6. Motion	No	N/A	
7. Defects	No	N/A	
Action plan			
Scope of project quality documentation needs to be addressed before a similar project is embarked on. In-house resource and capability for the technology required needs to be addressed through training planning before the next project.			

During delivery

At some stage in a project, it may become apparent that the project is out of control and in a state of chaos, either because of a single event or as a result of a project review. In either case, the Project Manager and his team need to react quickly and solve the root cause of the crisis. There are many root cause analysis techniques, but the one demonstrated here is the fishbone analysis. The Project Diagnosis Fishbone tool (Figure 5-17) can be used to brainstorm cause–effect relationships against standard project failure categories.

Post-delivery

When a project has failed, a formal analysis of the causes is an important review stage. Using root cause analysis is also appropriate in this case (Figure 5-17). Project failure can add value if the organization learns from it and puts processes in place which prevent a reoccurrence.

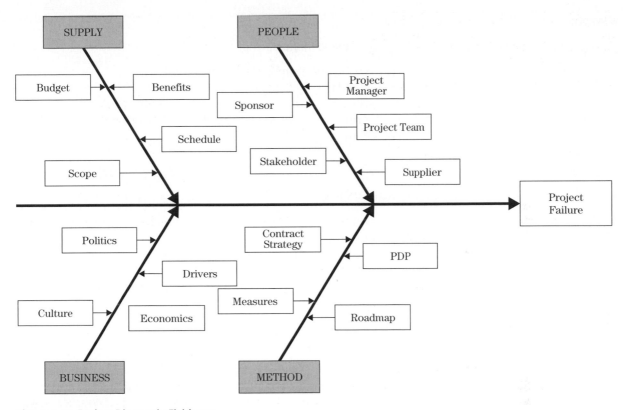

Figure 5-17 Project Diagnosis Fishbone

Tool: Project Diagnosis Fishbone

The Project Diagnosis Fishbone tool (Figure 5-17) is intended to help structure a project review. It is based on the Ishikawa diagram and aims to link cause and effect when considering a project failure (either one aspect as in a crisis situation or an overall project failure).

The approach is to divide the influences on the project, or causes of the crisis, into the four main groupings and then within each group indicate positive and negative influences on the outcome. The intention is to provide a structured approach to thinking and remove any emotional component from the analysis. Some typical questions that can be asked are:

Supply

How could the project requirements and/or constraints cause failure?

- *Budget* – Is the budget adequate for the scope?
- *Benefits* – What benefits should be delivered? Are they attainable?
- *Schedule* – Is the schedule realistic?
- *Scope* – Is the scope defined? Does it align with the schedule, budget and benefits?

People

How could the people involved in the project cause failure?

- *Sponsor* – Is there a sponsor who is engaged and supporting the project?
- *Stakeholders* – Are the stakeholders defined and engaged?
- *Project Manager* – Is the Project Manager in place and capable of delivering the project?
- *Project Team* – Is the Project Team identified, committed and capable of delivering the project?
- *Suppliers* – Are suppliers identified and are they performing?

Business

How could the business environment cause failure?

- *Politics* – Are there political pressures on the project? Do people or groups have their own agendas? This is a particular issue with change projects.
- *Drivers* – Have the business drivers changed since the project was started?
- *Economics* – Have financial conditions changed which may affect the project?
- *Culture* – Is the organizational culture affecting the way the project is being implemented?

Method

How could the way the project is delivered cause failure?

- *Contract strategy* – Is the strategy working with or against the project approach? For example, is the strategy to go for fixed price, inflexible contracts when the project is still defining the scope?
- *Project Delivery Plan (PDP)* – Is there a PDP which identifies the approach and is it consistent with the aims of the project?
- *Measures* – Are the project performance measures appropriate to success (on the basis that measures cause behaviours)?
- *Roadmap* – What is the project roadmap? Is the project delivery approach consistent with the roadmap?

Examples of the use of the Project Diagnosis Fishbone are shown in Chapter 6.

Control plan delivery case study – construction project

To illustrate the key points from this chapter, the delivery phase of a construction project is described. It focuses on the delivery of the control plan.

Situation

A new headquarters for an organization based in mainland Europe was due to be built in a number of phases:

- **Phase 1** – Infrastructure for the campus, including all temporary roads and facilities included in later phases.
- **Phase 2** – Offices for the corporate division and all marketing research departments.
- **Phase 3** – Offices for the various product divisions and conversion of all temporary facilities into final versions.

Phase 1 had proceeded to plan, and the organization was so satisfied with the outcome that they decided to use the same main contractor. However, Phases 2 and 3 were more complex in terms of the build structure and architecture. The time-scale was also a more significant challenge as people needed to be moved into the offices more quickly than originally expected due to a recent company expansion.

Project planning

For Phase 2 (to be quickly followed by Phase 3), the development of a control plan was particularly critical in managing a vast array of external suppliers, significant cost and schedule constraints and a large, complex construction site (aiming to achieve benchmark safety and quality metrics).

The Project Manager, with support from other project management expertise from within the company, completed the development of a thorough control plan. Due to the type of project, the majority of the Project Team consisted of external resource with a few internal facility management experts and a small 'internally resourced' project office (accountants, planner and administration). The control plan was based on the defined path of success (Figure 5-18) and focused on:

- **Risk management** – ensuring that risk was placed where it could be most appropriately managed.
- **Contract management** – ensuring that the multiple contracts were in control.
- **Control metrics** – ensuring that forecasting was based on robust data.

Project intervention

The Project Manager (Michel Noir) assigned to the Phase 2 project was inexperienced, so a more senior Project Manager (Sally Edwards) was asked to provide support by conducting regular project health checks. These checks were based around a review of key control metrics which indicated whether the project was delivering according to the control plan. Almost immediately issues appeared (Figure 5-19), so after the first health check a higher degree of future intervention was agreed.

<table>
<tr><td>

CSF 1
UNDERSTAND CUSTOMER NEEDS
Ensure that the completed facility is of the appropriate quality and operability to meet business needs as articulated by the customers

</td><td>

CSF 2
ROBUST PROJECT MANAGEMENT
Ensure that the project uses effective project management processes, remains in control and can accurately forecast the project outcome

</td><td>

CSF 3
APPROVED DESIGN
The design work is completed within budget and on time, and is subsequently approved for implementation on the basis that it reflects the appropriate facility quantity, quality and functionality requirements

</td></tr>
<tr><td>

CSF 4
SAFE AND TIMELY CONSTRUCTION
The construction work is completed within budget and on time, and delivers the appropriate facility quantity, quality and functionality requirements safely

</td><td>

CSF 5
EXCELLENT SUPPLIER PERFORMANCE
Ensure that the suppliers use effective project management processes to deliver the quantity, quality and functionality requirements

</td><td>

VISION OF SUCCESS
The Phase 2 facility is handed over to the business on time and within budget, meeting all business requirements. Customers are satisfied: from senior stakeholders to end-users working in the corporate division and marketing research department.

</td></tr>
</table>

Figure 5-18 Path of success – construction project

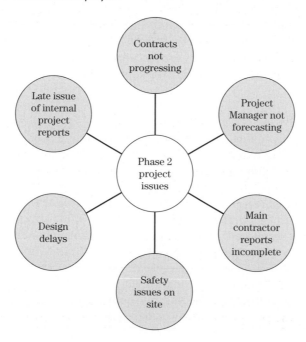

Figure 5-19 Phase 2 issues – construction project

Following the health check, Sally conducted a root cause analysis and concluded that:

➤ The Project Manager was not using the control plan and consequently was not in control. Because of this the main contractor was not being managed appropriately.
➤ The main contractor was not delivering its aspects of the control plan. Consequently other suppliers within its control were showing signs of future performance issues.
➤ The control plan was then reviewed more closely to ascertain whether any mitigating actions to solve the root cause were needed beyond that of a stronger intervention style.

Control plan delivery

Once the control plan was used, the Senior Project Manager was able to review progress more effectively and intervene as needed. The reviews focused on three basic areas:

- Risk profiles.
- Contract status.
- Cost, schedule and safety metrics.

Risk management

During the early part of the delivery when design, construction and contract placement were all in progress at the same time, the risk profile was tracked very closely. By the fifth review, the project trends were increasingly negative and further intervention was needed (Table 5-20).

Table 5-20 Project – Construction Project Risk Profile

Delivery Toolkit – Project Risk Profile						
Project: Construction project			**Sponsor:** Pierre Bleu			
Date: Month 7, year 1			**Project Manager:** Michel Noir			
Risk profile trend						
Risk area	**Milestone review dates**					
	Project start	**M3 1st review**	**M4 2nd review**	**M5 3rd review**	**M6 4th review**	**M7 5th review**
CSF 1 – Customer needs	Green	Green	Green	Green	Amber	Amber
CSF 2 – Project management	Green	Red	Amber	Amber	Red	Red
CSF 3 – Design	Green	Amber	Amber	Amber	Amber	Red
CSF 4 – Construction	Green	Amber	Amber	Amber	Amber	Red
CSF 5 – Supplier performance	Green	Red	Amber	Amber	Red	Red
Project risk summary						
Mitigation plans put in place after the 1st review are no longer working. The root causes need to be identified and resolved quickly otherwise project success will be compromised. Neither the Project Manager nor the main contractor can adequately explain the reasons behind the increasingly negative trend and the attitude appears to be that 'it's a project – things go wrong'.						

The initial intervention after the first health check appeared to have rectified many of the initial issues but as the project progressed more and more problems surfaced. In four CSF areas, risks had become issues, and an issue planning session identified the most appropriate action plan (Table 5-21). As a result, some critical changes were made:

- **The Project Manager was replaced.** In reviewing the root cause for many of the issues, it became apparent that the Project Manager could not cope with the size and complexity of the project. His lack of experience in contract management in particular was exposing the organization to significant financial risk.

Table 5-21 Issue Action Manager – construction project

Delivery Toolkit – Issue Action Manager							
Project:	Construction project			**Project Manager:**		Michel Noir	
Date:	Month 7, year 1			**Page**		1 of 1	
Issue log							
Issue number	**Issue description**	**Issue owner**	**Issue status**	**Issue action plan**		**Action owner**	**Action status**
1.1	Customers were unhappy with the design work they have seen	Pierre Bleu	Critical Live	The external design team are to work with the main contractor to develop a model of the office as it will look at handover		Michel Noir	New action
2.1	Forecast cost is 20% over budget	Michel Noir	Critical Live	Conduct a full review of the scope to see where the cost overruns are forecast. Reduce the overrun to 5% or less		Sally Edwards	Review meeting set up
2.2	Schedule is forecasting an undetermined overrun	Michel Noir	Critical Live	Conduct a schedule review event with representatives from the main contractor, design contractor and those suppliers who have already started on site		Sally Edwards	Review meeting set up
2.3	Project reports are inaccurate and driving poor decisions	Michel Noir	Critical Live	Design a lean report and implement		Sally Edwards	Proposal made
2.4	Project Manager is out of control and does not know what the outcome of the project will be	Pierre Bleu	Critical Live	Replace Project Manager and induct replacement into the role		Sally Edwards	Proposal made
3.1	Design drawings are not available in time to place relevant contracts	Michel Noir	Near Critical Live	Conduct a review of the design process in terms of efficiency and effectiveness		Design manager	New action
3.2	Design drawings are not of the right quality	Michel Noir	Near Critical Live	Root cause issues for poor quality and resolve within the next week		Design manager	New action
4.1	4 minor safety incidences on the contraction site last month	Michel Noir	Critical Live	Root cause issues for poor safety record and resolve. Start twice daily safety walk around		Site Manager	New action
4.2	Construction contracts are behind schedule	Michel Noir	Near Critical Live	Conduct a site review to identify root causes for the delays and resolve		Site Manager	New action
5.1	Contract plan is behind schedule	Michel Noir	Near Critical Live	Review the procurement process to identify root causes and resolve		Contract Manager	New action
5.2	Main contractor is out of control and not managing suppliers	Michel Noir	Critical Live	Have a 1-1 with the main contractor Project Manager and put in new controls		Sally Edwards	Meeting set up
Issue action summary							
Michel agreed that the project was too much for him and in fact the stress was making him ill. He wanted to be reassigned or receive more day to day support from Sally Edwards. The recommendation was to replace Michel and a replacement has been proposed by Sally. All of Michel's actions will be passed to the new Project Manager once he is in role.							

- **The main contractor was managed more closely.** The root cause for a number of poor decisions by the Project Manager was poor data received from the main contractor. As a result the new Project Manager, Georges Blanc, gave the main contractor the project management processes and tools which he expected them to use to report progress.
- **The project cost plan and schedule were reviewed.** As a result tasks were re-planned in order to meet the original goals which were business critical.
- **The customers were invited to view a 'sample office'.** Customers needed to be engaged in the design process so that they could be assured that their requirements would be met.

Georges, quickly got on board with the issues, and within 2 months control had been regained. This was demonstrated by the positive changes in contract status and project outcome forecasts. In addition, he conducted a waste review with Sally as an induction into the project practices and to also be sure that they were adding value (Table 5-22). This induction review underlined why some of the critical issues were occurring.

Georges and Sally agreed a new project-reporting process and format which comprised:

- The Project Risk Profile (Table 5-20).
- The Issue Action Manager (Table 5-21).
- The Contract Tracker (Table 5-23).
- The Project Control Tracker (Table 5-24).

Table 5-22 Project Waste Analysis Tool – construction project

Delivery Toolkit – Project Waste Analysis Tool			
Project:	Construction project	**Sponsor:**	Pierre Bleu
Date:	Month 8, year 1	**Project Manager:**	Georges Blanc
Team checklist			
Observed waste	**Response**	**Comment**	
1. Over production	Yes	Design teams are doing more work than is necessary to deliver documents to construction teams. The level of detail is not adding value and is contributing to delays	
2. Waiting	Yes	Contracts are not being placed as they all require sponsor authorization. This is not lean. In future, only those greater than €250,000 need his signature.	
3. Transport	No	All teams have co-located to site, including the design team (which is made up of representatives from all relevant external suppliers) so documents do not need to travel	
4. Inventory	No	Project documentation and processes to manage documentation are appropriate	
5. Over processing	Yes	The monthly progress report is a collation of individual reports is, condensed and reviewed many times before it gets to the sponsor. Critical detail is lost and it takes 3 weeks to produce. This is not lean and needs to be changed.	
6. Motion	No	No issues with Project Team travel – all co-located	
7. Defects	Yes	Design errors, site errors – all being resolved via critical action plans	
Action plan			
In general the project processes now in place are lean with a few exceptions. The issue with defects is a product of some over production and can be easily resolved. The reporting issues are indicative of the poor main contractor performance which is also being resolved.			

Table 5-23 Contract Tracker – construction project

Delivery Toolkit – Contract Tracker

Project:	Construction project		Project Manager:				Georges Blanc
Date:	Month 10 year 1		Page:				1 of 1

					Contract tracking		
Contract	Type	Supplier	Planned status	Actual status	Threat to project	Mitigating action
Construction management contract	Target reimbursable fees versus man-hour estimate	Main Contractor A	50%	60%	High	Using man-hours at a greater rate than forecast. This is due to their inefficiencies during early part of contract. Need to resolve this – possibly convert to fixed fee.
Design consultant	Target reimbursable fees versus a schedule of deliverables	Design Consultant B	100%	90% Earned value at 80%	High	Significant progress since intervention in Month 7. Cost variations due to their errors and inefficiencies already rejected.
Civil works contract	Bill of quantities	Contractor C	100%	80% Earned value at 79%	Medium	Final works will not prevent other contracts starting. Outcome cost remains in budget
Building shell sub-contracts (10)	Fixed price and agreed variation SOR	Various contractors	All contracts placed	80% contracts placed	Medium	All critical contracts now placed, with non critical ones near to completion. Site start in next 2 weeks for critical contracts.
Internal building sub-contracts (7)	Fixed price and agreed variation SOR	Various contractors	All contracts placed	57% contracts placed	Medium	Contracts with long-lead materials were prioritized so that schedule is not impacted.
Building infrastructure sub-contracts (9)	Fixed price	Various contractors	All tender packages issued	100% issued	Low	Ensure that each contract contributes to cost saving needed
External campus infrastructure sub-contracts (7)	Fixed price	Various contractors	All tender packages issued	95% issued	Low	Ensure that each contract contributes to cost saving needed

All contracts placed since Month 7 are contractually robust, within budget and on schedule. The mitigation focus has been to place all critical contracts so that the schedule and cost issues are resolved and the forecast at completion can be analyzed more accurately. Prices negotiated on new contracts to date have saved 6% from budget with a further 5% estimated from future contracts. Variations to older contracts have been reduced or rejected saving a further 5%.

Table 5-24 Project Control Tracker – construction project

Delivery Toolkit – Project Control Tracker					
Project:	Construction project		**Project Manager:**	Georges Blanc	
Date:	Month 10 year 1		**Page:**	1 of 1	
Cost					
Area	**Control tool**	**Current status**		**Action/mitigation and forecast status**	
Spend as planned	Cost plan	Estimated cost overrun has decreased from 20% (M7) to 4%	AMBER	Current mitigation via contract development and variation management is working	GREEN
Spend un planned	Cash-flow profile	Cash flow is still behind schedule	AMBER	Maintain focus on contract placement (Figure 5-20)	GREEN
Spend risk	Contingency spend	Contingency is fully allocated with an estimated 4% overspend	RED	Maintain focus on cost savings and risk mitigation so that no contingent funds are needed at project end	AMBER
Schedule					
Area	**Control tool**	**Current status**		**Action/mitigation and forecast status**	
Done when planned	EVA	Project schedule delays in design and procurement have been eliminated	GREEN	Maintain focus on earned value progress management	GREEN
Done by who planned	Resource plan	Resources have now settled down after a purge in the design team	GREEN	Maintain focus on quality management	GREEN
Schedule risk	Schedule buffer	Project buffer of 1 month is fully allocated	AMBER	Maintain focus on contract placement, site management and schedule risk mitigation	AMBER
Change					
Area	**Control tool**	**Current status**		**Action/mitigation and forecast status**	
Staff turnover	Staff turnover control chart	Numbers on site have increased and quality and safety measures have lead to requests for supplier staff to be replaced	AMBER	Maintain current overview on supplier staff issues	GREEN
Document revisions	Document revisions	Design documents are issued for construction within 4 revisions	GREEN	None	GREEN
Control change	Change log	Changes approved to reduce cost, meet schedule and customer needs	GREEN	Maintain current process	GREEN
Project control summary					
Forecast at project end is a 3% overspend whilst meeting schedule and customer goals. The internal project management costs are not currently attributed to the project, however the total man-hours forecast will be doubled due to intervention strategies needed to meet goals. No further safety issues have been seen and the quality of the build on site is excellent. The sample office achieved its goals in allowing end users to test their new environment. This allowed the Project Team to successfully test some less expensive options for building management (lights, air conditioning) and office furniture.					

Contract management

Georges recognized that much of his time was spent managing the main contractor and through them the external suppliers. However, the number and complexity of contracts was too critical for him not to be 'hands on' particularly considering the poor performance of this contractor to date. By Month 10 the Contract Tracker (Table 5-23) was showing significant progress, and the project was back in control.

Control metrics

Georges used a Project Control Tracker to summarize the control status on one page (Table 5-24).

This helped the sponsor to understand the current situation and so communicate this effectively to the customer group. It also helped the smooth running of the project health checks being conducted by Sally Edwards. These continued each month for the first 6 months after Georges was appointed as the Project Manager and then moved to quarterly.

Additional control tools, such as tracking charts (Figure 5-20) were used and displayed in the project office on site. These showed key metrics which the team could alter through their performance:

➦ **Contract placement** (number of contracts placed to date versus plan) – Schedule adherence depended on this control metric achieving plan.
➦ **Construction man-hours** (total man-hours on site versus plan) – This metric looks at schedule and safety. It is easy to try to catch up schedule on a construction site by starting more contracts at the same time. However, this can only be done after a safety review in terms of where the people are working and the type of work they are to do.

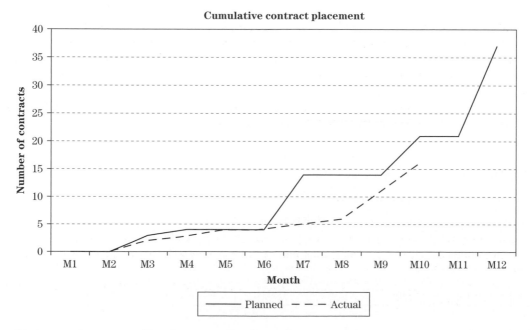

Figure 5-20 Contract tracking chart – construction project

- **Safe construction man-hours** (total man-hours since the last incident) – This is a safety metric used to motivate contractors to be safe and conduct safe working practices.

In addition, the board contained a marked-up copy of the campus layout, showing who was working in which area that week.

Project-cost tracking was confidential between the Project Manager, Senior Project Manager and sponsor, and various metrics were used to ensure control:

- *Project cost plan adherence* – used cumulative cash flow tracking tools to forecast total project cost outcome.
- *Cost plan adherence per area* – this metric looked at sets of similar contracts to see if cost performance was better or worse in specific areas of the works.
- *Contract variations* – number and total cost (approved, rejected and in dispute).

The project continued to be in control until it was completed on schedule.

Conclusions

Although Phase 2 of the project went through an extremely 'uncontrolled' stage, this was in part due to the lack of use of the control plan by the Project Manager, requiring intervention. The regular independent review conducted by a Senior Project Manager from within the organization identified at a very early stage in construction that progress was behind schedule, costs were escalating and external suppliers were not adequately coordinated or controlled. The root cause was poor project management at the overall level and at the supplier level.

As a result, a new Project Manager was appointed, and all parts of the control plan were implemented, bringing the project back into control. Phase 2 finished on time with an 8% cost overspend, no major safety incidences and a satisfied customer (the employees who were using the campus).

As a result of the issues during Phase 2, Phase 3 kicked off with a stronger project management team and a different main contractor. The change of contractor was necessary due to the project management issues which were never entirely resolved during Phase 2. They were only controlled through heavy intervention and detailed task management. In effect, the organization was paying for project management *and* then having to do it!

Phase 3 was planned and delivered successfully and the organization learnt a lot more about project management, which it was able to use in future business projects.

Key points

The aim of using this case study was to demonstrate that:

- It is no good having a control plan unless it is used.
- There is always value in having an independent project review to highlight the issues which a Project Manager can be 'blind to' when in the middle of a complex fast-track project.
- Delivery of the control plan needs to cascade into the supplier contracts so that contract delivery fully aligns with the overall project requirements

Troubleshooting control plan delivery

Table 5-25 is a list of common delivery issues associated with control plan delivery.

Table 5-25 Troubleshooting control plan delivery

Typical symptoms	Example root causes	Example solutions
1. Not following the control plan	The Project Manager isn't fully engaged in the project	Ensure that the Project Manager has the time and commitment to deliver the project
2. Poor supplier performance	Inappropriate supplier selection (however, this is more about planning)	Choose to either manage the contractual relationship or get out of it. Choice may depend on the contract status – if it's early then cancelling the contract may be more appropriate.
	Poor management of the supplier	Ensure that the interdependencies between the Project Team and the supplier are understood and managed
	Poor or no supplier engagement with the project goals	All major suppliers should be made aware of their impact on the ultimate success of the project
	Contractual issues or disputes	Ensure that there is a clear relationship between the supplier and the Project Team. Each supplier should have a named contact person (nominated from the Project Team) who manages the supplier as a key stakeholder.
	Supplier uses a chain of sub-contractors – risk, contract, performance	Ensure that the contract and sub-contract strategies are robust and aligned to delivering the project goals
	Supplier can't do what he said he would	Contract needs to be stopped, reviewed and then cancelled as necessary
3. Supplier drives schedule	Supplier financial concerns (recognizing sales before year end)	Understand the impact of project decisions on supplier business plans
4. Risks all occur	No risk mitigation	Define and deliver mitigation plans. Review risks on a frequent basis as part of a formal risk management process.
5. Risks occur and have major impact	No contingency plans in place	Define contingency plans for all high risks and when these should be enacted. Review risks on a frequent basis as part of a formal risk management process.
6. Significant increase/decrease in risk	Mitigation plans not working (increase in risk)	Review risks and evaluate new mitigation strategies
	Mitigation plans working (decrease in risk)	Maintain mitigating actions and keep monitoring the risk
7. Risk profile is negative	Starting a high risk project	Ensure that the sponsor and other key stakeholders understand the true status of the risk – communicate this frequently/effectively

(Continued)

Table 5-25 (Continued)

Typical symptoms	Example root causes	Example solutions
8. Fire fighting	No risk prioritization – just dealing with the next risk seen	Use risk techniques to ensure that resources are allocated to the highest risks
9. Uncontrolled change (scope, cost, time)	Limits of control are not defined within the Project Team	Ensure that formal change control procedures are in place and in use
	Change control procedures not in place or being used ineffectively	
10. Behind schedule	Unrealistic schedule	Review the schedule with the team and re-baseline. Communicate all changes to appropriate stakeholders after gaining buy in from sponsor.
	No schedule management	Ensure that progress measurement and management occurs at all levels in the Project Team
	Intended schedule delay	Ensure that this is agreed with the sponsor and any other stakeholders and then appropriately communicated
11. Over budget	Unrealistic cost budget	Review the cost plan with the team and re-baseline. Communicate all changes to appropriate stakeholders after gaining approval from sponsor and others as appropriate to the funding approval procedures within the company.
	No cost management	Ensure that cost spend and management occurs at all levels in the Project Team
	Intended cost overrun	Ensure that this is approved by the sponsor and any other stakeholders and then appropriately communicated
12. Project spend appears less than it really is	Historic lag in accounting results (last months figures by end of this month) – the data from systems is delivered too late to be of any use in 'live' control	Need to understand if there is a lag in the system and have an appropriate method to forecast costs accurately. Link project cost and finance systems – don't rely on any one.
13. The last 10% of progress takes 90% of the time	Unrealistic progress measurement – not really at 90% complete	Ensure that the progress measurement system is linked to a form of earned value rather than subjective assessment
	Progress measurement system is inappropriate – can't measure the task accurately	Ensure that the progress measurement system can discriminate stages in the task – identified earned value
	Loss of focus as commence next task	Recognize when resources are being asked to multitask and consider prioritization and appropriate communication of targets – keep the last task on the radar
14. No learnings is captured	No time for project reviews	Conduct reviews at the end of each key project stage – use a focused AAR technique to manage learning and knowledge management
15. Project Team blamed for failure	Lack of communication that the project was going wrong	Project Manager needs to communicate potential failure with the sponsor and either mitigate failure or agree the appropriate way to communicate it
16. End of project blues	Lack of communication and associated recognition of project success	Project Manager needs to communicate success with the sponsor and team and agree the appropriate way to communicate and reward it

Handy hints

No control plan means no 'control'

If you don't have a plan, then you have no baseline from which to check progress, assess change or forecast from.

Control isn't about rigidity

Control is about flexibly managing a situation to achieve a specific goal. After all, not all change is a bad thing if it supports ultimate project success.

It's OK to be a control freak

If you know where you're going and when and how you'll get there, then that's something to be proud of.

If you're not forecasting you're not controlling

If you are just monitoring what is happening, that's history. If, on the other hand, you take that history and use it to assess the future, you are starting to control the world around you (your project).

Mitigate risks and manage issues

A risk hasn't yet happened, and we want to make sure it doesn't happen. But if it does, we need to manage the subsequent issue for the project.

Stop reinventing the project wheel

Every project is an opportunity to learn. If project reviews aren't conducted or actions to share knowledge aren't completed, then we are just wasting all the good stuff we have gained from a project. We will only *not* make the same mistakes next time because we knew why they happened.

And finally . . .

Effective control plan delivery:

- Maintains the link between cost, scope and time to manage uncertainty and forecast the project end-point.
- Supports business plan delivery, ensuring that the business benefits will be enabled.
- Supports set-up plan delivery, ensuring that the Project Team deliver value, which supports project success.
- Is built on robust project process delivery and effective project control management.

6 When projects go wrong

When projects go wrong, the impact is felt at many levels. The organizational impact of the following is well documented and was covered in *Project Management Toolkit* (Melton, 2007):

- Delivering a project late.
- Delivering a project over budget.
- Delivering a project which does not meet scope requirements.

However, project success is not determined by these three basic parameters alone. Organizations use projects as one mechanism to deliver business goals, and so project success can also be analyzed by success in achieving those objectives (Melton et al., 2008). Figure 6-1 summarizes how the concepts in this book lead to a definition of project success.

Project objectives		Business objectives	
		Missed	**Met**
Met		**White elephant** Delivery of the approved PDP but no link to the business Likely robust project controls but no delivery of sustainable benefits for the business	**Star** A project which delivers the approved business case and delivers sustainable business benefits Effective project management
Missed		**Failure** Non-delivery of the business case and PDP Likely poor project management coupled with ineffective sponsorship, so no robust link to the business	**Question mark** Delivers sustainable benefits for the business but likely that business case and PDP are not delivered completely A multitude of root causes

Figure 6-1 Project success matrix

This four-quadrant model allows a categorization of project outcomes:

Failure

These are projects that do not deliver either project or business objectives. Case studies A, B and C demonstrate the root causes of some project failures. Common to all three examples is the fact that aspects of poor project management practices were demonstrated.

White Elephant

These are apparently successful projects which ultimately do not deliver intended business benefits. Case study D demonstrates sample root causes of this situation. Although there are many root causes for this type of outcome a common one is the lack of linkage between the project and the business. For example a poor relationship between the sponsor and the Project Manager or an inadequate business case delivering an inappropriate PDP.

Question Mark

These are projects that have had internal problems during delivery but which ultimately meet the business needs even if project objectives are not delivered. Case study E demonstrates sample root causes of this situation. Often this type of outcome is labelled a failure at project close-out. The challenge is to identify the ultimate outcome (business benefits delivery) so that the project team do not spend too much time analysing their apparent failure.

Star

These are projects that meet and/or exceed both project and business objectives. The case studies outlined in Chapters 7 and 8 illustrate are in this category. They demonstrate use of good project management practices.

Case study A – business change project

It is often those projects which are undertaken purely to change the business that fail, whether this is recognized or not. Often business changes are not set up as projects and so suffer from poor project management practices in every stage of planning and delivery. They also generally only solve high-level symptoms of problems and this very rarely leads to sustainable business success. Case study A highlights this type of ambiguous situation.

Situation

The quality assurance laboratories (labs) for a growing production unit within a major contract chemicals manufacturing organization were identified as a bottleneck, inhibiting the company's ability to increase the overall plant capacity:

- Samples took extended periods to test (over and above test time targets within their sales and planning (SAP) system).
- Samples or associated paperwork was often mislaid, further extending test time.
- Analysts worked a normal working day rather than production hours (currently an extended 2 shift 16 hour working day but due to go to 24-hour working in the near future).

The QA Manager had repeatedly asked for funds to either buy new test equipment or to completely refurbish the labs. In the light of the proposed capacity expansion, he had also asked for additional staff and a new lab for them to work in.

However, the QA Director (who had overall responsibility for all European sites) was not satisfied that the lab was maximizing its current capacity. He would not support any changes which required more capital expenditure (CAPEX) or operational expenditure (OPEX) until he was satisfied that this was the case.

The QA Manager had started a review project internally but this faltered due to the high workload of his team. He therefore asked for support from the central change team within the organization.

Pre-delivery

A small team of change managers came to site to conduct a project assessment using their business improvement processes and tools. They observed current ways of working although they made no attempt to engage with the analysts (on instruction from the QA Manager). Their initial view was that:

- Samples entering the lab were not dealt with in any systematic manner.
- Each analyst had 'ownership' of a particular sample for all required tests, leading to inefficient use of complex and expensive test equipment.
- At times, there were queues for equipment and at other times equipment was idle for most of the day.
- Samples generally queued up after the analysts left for the day as the production area was still operational.
- The QA Manager had no formal method of status assessment other than an ad hoc daily walk around.

Based on this, the change team developed a project charter which was agreed with the QA Manager.

Delivery

The QA Manager made it clear from the start that there was to be no interruption of the analysts' work. He was already in trouble with the QA and Production Directors due to the delays testing was causing and did not want to compound this. Therefore, the lead change manager enlisted other people in the change team to work on the project to collect data, design new ways of working and then implement these.

The data very clearly indicated that the lab could operate at 50% higher capacity through changing specific ways of working:

- Develop an extended day shift system so that the lab was in operation during more of the production time.
- Develop a test scheduling system to track tests as they enter the lab, as they are tested and as they leave.
- Split the lab up into test zones and have analysts assigned to that zone on a rotating basis – thus optimizing equipment and analyst time.
- Use a pigeon hole storage system to hold all sample paperwork and also act as a visual method of tracking sample status.

The design was presented to the lab teams immediately before implementation and caused much concern, particularly amongst the senior analysts. Due to the teams' resistance, only half the project objectives were delivered. However, benefits tracking showed that even this part-implementation had increased capacity by 25%. The change team left the QA Manager to complete the work based on a list of outstanding project activities including the change in working hours, changes to lab layout and the visual tracking of samples and paperwork. These were never completed.

Post-delivery

The changes in ways of working which delivered the 25% capacity increase required the analysts to work in small teams to complete specific tests on various samples. However, as soon as the change team left, the senior analysts returned to their previous ways of working saying they were unable to fully utilize their skills by working in a 'production cell'. Within a matter of weeks capacity was back to pre-project levels and production was once again impacted, ultimately impacting customer delivery lead times.

The project did not achieve either the project or business objectives and was considered a complete failure. However, the QA Director had observed the short spike in capacity and remained convinced that performance could be improved. He felt sustained capacity improvements could still be made if the project were to be revisited and redeveloped. He insisted on a full project diagnosis as a way to develop the follow-on project.

Root cause

The principal causes of this project failure are shown in Figure 6-2, and Table 6-1 contains more detail of the root causes for each of these.

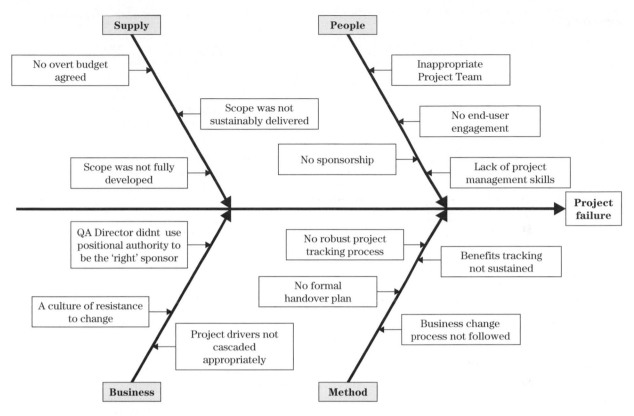

Figure 6-2 Diagnosis Fishbone business change project failure

Table 6-1 Assessment of failure – case study A (business change project)

Failure area	Root cause	Assessment
People	No end-user engagement	The QA Team was not involved at all. Their ideas and feedback were not requested and communication to them on the status of the project was minimal.
	Inappropriate Project Team	The Project Team did not include any of the QA Team. Bearing in mind that this project was all about changing current ways of working and developing and testing new ones, this inevitably lead to problems: incorrect assumptions, inappropriate solutions (or ways to implement solutions) and further alienation of an already hostile group.
	Lack of project management skills	The change managers involved in this project had a vast experience in the use of change tools but had little experience in managing complex changes from start to end. As a result the delivery methodology was flawed from the start.
	Lack of sponsorship	The QA Manager took on the sponsor role but he was not encouraged to take an active role. He didn't support communication either within his QA Team or externally to the QA and Production Directors. This lead to an overall lack of stakeholder management.

(Continued)

Table 6-1 (Continued)

Failure area	Root cause	Assessment
Method	No robust project tracking process	The project scope was not tracked formally in terms of what had been delivered and if this was on schedule or within budget
	No formal handover plan	At handover, a list of outstanding items was issued to the QA Manager but their importance to the project outcome was not discussed. In effect, the project was handed over before it was finished.
	Benefits tracking not sustained	The benefits were tracked by the Project Team. Overall capacity and test cycle–time tracking commenced early in the project to get a baseline; however, the process should have been handed over so that the tracking was sustained.
	Business change process not followed	Although the change managers were experienced in the use of business improvement tools, they were not experienced in the complete change cycle considering the soft and hard aspects. They failed to consider the readiness for change in this environment.
Supply	No overt budget agreed	The changes required investment; however, there was no formal funding request. Therefore aspects of the design were simply not implemented, even though these were fundamental to the sustainability of the changes and consequently the benefits. For example, the cell-based working required some lab layout modifications which would ultimately have made it difficult for the analysts to return to their previous ways of working. This required capital funding and also a commitment to a short lab shutdown.
	Scope was not fully developed	Areas of scope were not fully developed and therefore not fully implemented. For example, the extension of working hours or use of shift working needed the early involvement of the HR department, but this was never within the scope.
	Scope not sustainably delivered	The scope was not delivered in a way which would ensure that the changes would be maintained after the project was complete. This was partly about what scope was delivered and partly about how it was delivered. For example, the analysts were best placed to group types of tests into cells based on test equipment used and specific skills needed. However, they were not asked and in the end, the grouping of tests was not optimal.
Business	QA Director didn't use positional authority to be the 'right' sponsor	This project needed the sponsorship of someone with sufficient authority to make the changes happen. However, the QA Director wasn't even aware that the project was in progress.
	A culture of resistance to change	The analysts were not alone in their attitude to change. The company environment was not used to large changes and was slow in reacting to change in the competitive environment.
	Project drivers not cascaded appropriately	The need for change, and therefore the need for the project to succeed, was not cascaded throughout the production and QA Teams

Learning points

This project illustrates some important aspects of project delivery: appropriate project roadmap and scope, robust business change management, appropriate team capability and effective sponsorship. Taking each in turn:

Appropriate roadmap and scope

Every project has a distinct roadmap. A Project Manager must define the appropriate roadmap and manage the scope so that it fits within it. In this project, the business change roadmap was not managed and key decision points which support business change success were not completed. In this project, key stage gates around design review, approval and pre-implementation were not used and as a result the implementation was always at risk.

Using tools such as the Roadmap Decision Matrix and Scope Definition Checklist from *Real Project Planning* (Melton, 2008) and the Scope Tracker (Table 4-7) support the definition and delivery of scope.

Robust business change management

Every project delivers a change to the business and every business change is a project. In order to be successful, the change needs to be sustained within the business so that it is no longer a part of the project scope, but integrated into business as usual. This integration needs all levels of stakeholder management, good communication and an understanding of the risks to change sustainability.

Using tools such as the Sustainability Plan from *Real Project Planning* (Melton, 2008) and the Business Change Mitigation Matrix (Table 3-16) support the effective delivery of business change sustainability.

Appropriate team capability

Each project has its own needs in terms of:

- The combination of project and technical skills.
- The knowledge and experience of similar projects or working environments.

When considering business change projects, a key capability is the knowledge of existing ways of working and the experience of operating within the specific business environment.

Using a tool like the Project Team Audit Tool (Table 4-5) supports the ongoing review of team effectiveness and the identification and mitigation of any capability issues.

Effective sponsorship

Without an active sponsor, there is really no organizational support for a project and certainly no robust link between the needs of the business and the project being delivered. This project did not have any real sponsorship, and the project was being delivered in spite of business needs rather than in support of them.

Using a tool such as the Business Plan Review Checklist (Table 3-4) helps to identify and manage sponsor, benefits management and business change management issues early in project delivery.

Case study B – IT project

Projects rarely fail completely, but there are occasions when scope, costs and control issues combine so that there is a failure to deliver both the project and the business objectives. Case study B illustrates this type of project.

Situation

The project was to develop and implement a data mining solution to support the business in analysis, investigation and decision making. The company was a manufacturer producing a number of different products for the consumer market. The rationale for the development of this data mining tool was a number of expensive product failures where unforeseen combinations of raw materials, intermediates and manufacturing steps had resulted in scrapped or reworked batches.

The concept of a data mining tool to help the business was originally proposed some years ago as a way to address the difficulties of accessing information from multiple sources (in-house business system, supplier information, processing conditions and analytical results). It was felt that such a tool would help consolidate the information into a single location to enable easier investigations.

A project was eventually proposed on the basis of this original concept with the emphasis on access to data from the business system. A prototype was developed to prove the concept and, after a brief review, the project was fully sanctioned. The implementation was contracted out to a third party supplier.

Implementation of the project proceeded for approximately 9 months at a cost of £1.4 million before any formal review of the output. At this point, it was realized that the project was not going to meet the needs of the users. Essentially, the solution would have required significant manual input of data. It was therefore decided to terminate this phase of the project and review an in-house approach with a focus on adding additional functionality to satisfy the users.

The redeveloped project was re-sanctioned with a slightly revised scope using a combined prototyping/implementation model. In essence, this was a repeat of the initial project with an in-house team established to better reflect the user requirements. This second phase of the project took a further 9 months and incurred costs of approximately £1.5 million to develop and deliver the user requirements. However, it was also ultimately stopped once it was realized that it would not have been possible to consistently and reliably populate the system with the necessary data.

In summary, the project lasted 18 months and incurred costs of £2.9 million through a combination of external and in-house activities. The project was finally terminated when it was clear that the solution would not meet the business objectives.

Root cause

The principal causes of this project failure are shown in Figure 6-3, and Table 6-2 contains more detail of the root causes for each of these.

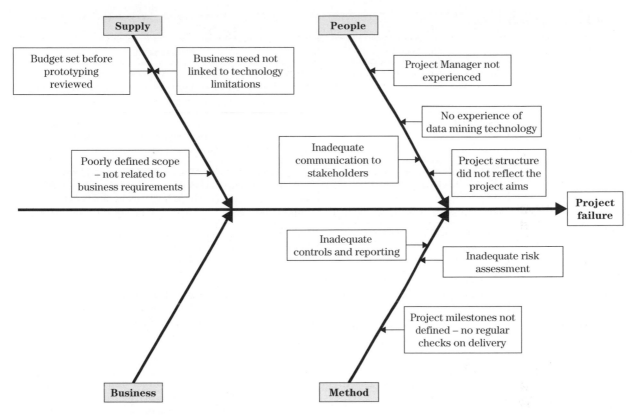

Figure 6-3 Diagnosis Fishbone IT project failure

Table 6-2 Assessment of failure – case study B (IT project)

Failure area	Root cause	Assessment
People	Project Manager not experienced	The Project Manager had no experience of working on projects with a high degree of manufacturing input requirement
	Inadequate communication to stakeholders	Stakeholders (end users and fund holders) were not adequately informed of the progress of the project in terms of solution capability and overall cost
	No experience of data mining technology	The Project Team had no experience of this type of technology and therefore had to 'buy-in' expensive expertise and learn on the job
	Project structure did not reflect the project aims	The project was structured around the needs of IT technology delivery and did not contain adequate representation from the user community

(Continued)

Table 6-2 (Continued)

Failure area	Root cause	Assessment
Method	Inadequate controls and reporting	Project costs were not actively managed. The Project Manager permitted spending without proper governance of progress, resulting in significant levels of spend.
	Inadequate risk assessment	Critical areas of risk within the project (ability to extract data from multiple systems and verification of data) were not identified and therefore could not be managed
	Project milestones not defined – no regular checks on delivery	The project milestones were established around hard deliverables (functional documentation for example) and were not granular enough to challenge the ability of the project to deliver
Supply	Budget set before prototyping completed	Prototyping and project delivery were rolled into one to facilitate approvals. As a consequence, the prototype was not adequately challenged before the main project was started.
	Business need not linked to technology limitations	The business wanted a solution which would deliver automated capture of all data. It was technically impossible to do this and the project (in its defined form) could never deliver the value to the business. There was no clear process to identify this and to stop the project. Project controls tended to focus on delivery of tangible products (specifications) without challenging the benefits delivery.
	Poorly defined scope – not linked to business requirements	The scope emphasized the 'easy to get' data whilst ignoring other, critical, data. The consequence was that the scope did not reflect the value the business actually sought in the solution.

Learning points

This project illustrates three important aspects of project delivery: adequate definition of the scope, robust engagement with stakeholders/Project Team and having project controls in-place and in-use. Taking each in turn:

Adequate definition of scope

Understanding what the scope was and how it linked to the business case would have highlighted the emphasis on data from the business system and the lack of linkage with the capability of current technology.

Using tools such as the Scope Definition Checklist from *Real Project Planning* (Melton, 2008) and the Table of Critical Success Factors from *Project Management Toolkit* (Melton, 2007) would have helped highlight these risk areas at the planning stage. Use of the Scope Tracker (Table 4-7) during delivery would also have highlighted that the deliverables would not meet the project goals.

Robust engagement with stakeholders/Project Team

Stakeholder and Project Team engagement are critical to the success of any project. In this case, the lack of communication with users and budget holders meant that the project failed to deliver on its business objectives, and there was no opportunity to terminate the project early. Furthermore, the team lacked the capability to deliver the scope and a better level of engagement would have highlighted this at an early stage.

Using tools such as the Stakeholder Engagement Tracker (Table 3-14) and the Project Team Audit Tool (Table 4-5) would have enabled better control of the people aspect of the project.

Project controls in-place and in-use

Unused project controls are more harmful to a project than no controls at all. At least in the latter, it is clear that the project is out of control! In this project, the Project Manager did not have adequate expertise to make proper use of the controls that were in place. The failure to link project and business needs meant that risk control was never put in place either.

Using tools such as the Risk Management Strategy Checklist in *Real Project Planning* (Melton, 2008) and the Project Risk Profile (Table 5-2) would have helped establish an understanding of the risks to project delivery and therefore better manage the uncertainty. In addition, the project tracking tools developed in Chapter 5 would have helped to establish a project control methodology.

Case study C – product launch project

Product development projects that deliver a product to market on time are seen as successful at launch. However, it is only when marketplace reaction is gauged that true success can be established. A project that seems successful at product launch can still fail in two distinct ways:

➤ Project failure – product does not work.
➤ Business failure – product does not sell.

Case study C is an example of an apparently successful product launch project failing in both ways.

Situation

At the time the project was initiated, there was market demand for a concentrated clothes washing liquid, packaged in smaller containers, that allowed the consumer to use less and still achieve the same or better washing results. This was at a time when concentrated cleaning products were seeing strong consumer interest. Liquid washing agents also offered some manufacturing benefits by eliminating the spray-drying stage of traditional powder manufacture.

The manufacturer developed a clothes washing liquid which met the criteria of improved cleaning capability in a smaller and more concentrated form. At the time, liquid washing agents were a novelty in the UK. A project was initiated to build a manufacturing and filling line for the new product, and a pilot launch was planned for a region of the UK.

When the product was launched in the market, there was a rapid and unforeseen impact. The washing liquid was widely reported as damaging fabric in 'spots'. This was essentially caused by the concentrate coming into direct contact with the clothes and small amounts of water just as the wash cycle started. Once diluted in larger volumes of water, the liquid performed well.

On investigation, it was clear that the way in which the washing liquid was used by the consumer was not in line with the trial activities carried out during product development. The negative publicity for the product was sufficient to cause it to be withdrawn and the project was terminated as an embarrassing failure.

Root cause

The principle causes of this project failure are shown in Figure 6-4, and Table 6-3 contains more detail of the root causes for each of these.

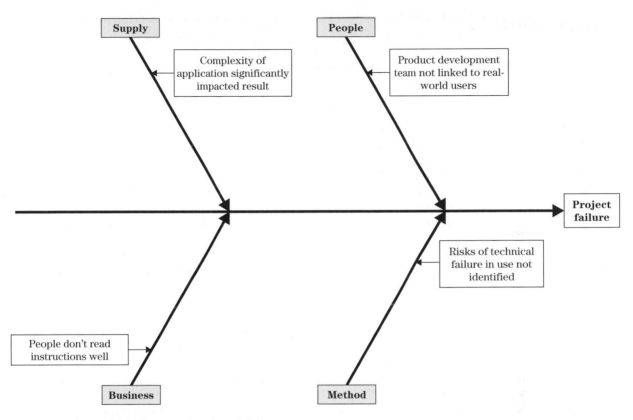

Figure 6-4 Diagnosis Fishbone product launch failure

Table 6-3 Assessment of failure – case study C (product launch project)

Failure area	Root cause	Assessment
People	Product development team not linked to real-world users	The product was being developed as a technical solution to a problem (a more concentrated and powerful detergent). There was no real engagement by the development team with real-world consumers who would not necessarily be 'optimum' users (i.e. use the product in the designed way every time).
Method	Risks of technical failure in use not identified	Failure of the product in use was not considered as a project risk. The effort was focussed on the risk of not being accepted by the target market (liquids were novel in the market in the UK at the time).
Supply	Complexity of application significantly impacted result	The liquid could not be introduced into the soap dispenser draw (it tended to run out) and a separate 'ball' was required, which needed to be put into the machine. If this leaked at the wrong point, the clothes could suffer damage.
Business	People don't read instructions well	Consumers favoured error free operation and tended not to read the instructions properly. This meant that the consumer was highly likely to introduce errors such as putting the detergent dispenser in upside down or pouring liquid directly onto the clothes.

Learning points

This project illustrates an important aspect of delivery, ensuring that the end-users are properly engaged and that the all project 'technical' risks are identified. Taking each in turn.

End-user engagement

Product development focussed on meeting the technical requirements of the project (cleaning capability in a concentrated package) and failed to adequately include the end-users (consumers). Generally, consumers are a non-compliant group and fail to follow instructions, particularly when they feel they either already understand the product (washing detergent) or don't believe it is significantly different from what they are used to using. Having sample end-users more engaged with the product development team would have identified the risks of misuse.

Using a tool such as the Stakeholder Engagement Tracker (Table 3-14) helps to focus on what the project needs from each stakeholder and whether they are getting it. It allows for action development when stakeholder engagement is too low and for when particular stakeholder risks are high, such as in this case. The ultimate success or failure of this project was always highly dependent on the buying and using behaviours of the consumer group.

Technical risk identification

Risk identification was essentially focussed on the risks to delivery of the new product for launch and its potential to fail (mainly linked to achieving the specified launch date). The product development team was not directed to look at product application issues and as such missed a huge technical risk: 'what if the product doesn't actually work?'.

Using the risk management tools in *Project Management Toolkit* (Melton, 2007) and *Real Project Planning* (Melton, 2008) to evaluate the risk to delivery and incorporating end-user risks, would be appropriate ways to track project issues prior to launch. Once the right risk categories have been developed, the Project Risk Profile (Table 5-2) can be used to track the progress of risks (and the overall risk profile) as the project is delivered. The starting point, though, for any risk assessment technique is to assess the critical failure points and so use of the Path of CSFs introduced in *Project Management Toolkit* (Melton, 2007) would have identified product technical issues such as consumer application problems, as critical to overall success of the product.

Case study D – biopharma facility project

When a project is delivered successfully (all success criteria or objectives within the PDP and stated business case have been successfully met), it is rare for project stakeholders to admit to the delivery of a white elephant. However, there are many examples of apparently successful projects which have delivered no benefit to their organization. Case study D is used as it highlights this unusual situation.

Situation

A pharmaceutical manufacturer had identified a business opportunity in developing and marketing biopharmaceuticals (biopharma). It had limited experience of this type of drug but was developing a licensing agreement with a small biopharma research and development (R&D) company. Based on the projected manufacturing volumes it soon became clear that the current pilot plant (based at the R&D company) would not be adequate for either market or regulatory needs, and so the building of a new biopharma facility was approved.

Pre-delivery

An experienced Project Manager and engineering Project Team was pulled together. Recognizing the need for biopharma technical knowledge, the Project Manager requested that scientists from the R&D company be released for the design and commissioning periods. A preliminary scope was developed and costed, and then a schedule pulled together. This was integrated into the funding approval request and based on market forecasts the investment payback was deemed acceptable.

Delivery

The project delivery proceeded as per plan with the engineers and scientists providing expertise where it was needed. Delivery was completed on time (to the requested handover date) and only 3% above budget. The Project Team used the previous three biopharma products as the base process upon which to test the facility and commissioning demonstrated performance to functional requirements. The project was hailed as a great success and it was an example to the company of the value of good project management in partnership with technical excellence.

Post-delivery

The facility remained unused until the first drug was ready for scale-up. Three months following formal handover to the production team, the R&D company announced that they had the first candidate drug which had undergone initial clinical trials. The drug manufacturing process was complex and for the first time a solvent (flammable liquids) was required to extract and purify the drug. Unfortunately, the facility had been designed around completely aqueous (water-based) processes and retrofitting it to safely handle flammable substances was not viable as a short-term option. The original pilot plant was therefore used on the basis that this was likely a one-off occurrence. However, the second and third candidate drugs also required a similar processing environment. After 2 years of waiting for a suitable process and after making alternative production arrangements for biopharma production, the facility was decommissioned and eventually sold. It had become a white elephant.

Root cause

The principal causes of this project failure are shown in Figure 6-5, and Table 6-4 contains more detail of the root causes for each of these.

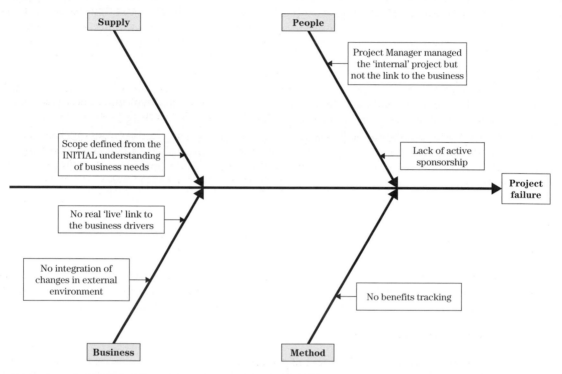

Figure 6-5 Diagnosis Fishbone biopharma facility project failure

Table 6-4 Assessment of failure – case study D (biopharma facility project)

Failure area	Root cause	Assessment
People	Project Manager managed the 'internal' project but not the link to the business	The Project Manager had a large, complex project to manage with significant technical hurdles to overcome due to the nature of the facility. However, such a single-minded focus on delivering cost, scope and time without looking at the business in which the project sits is not best practice and in this case was a contributory factor to an expensive white elephant.
	Lack of active sponsorship	The sponsor knew he had an experienced Project Manager and basically left him to it. He also wasn't actively managing the links between what the project would deliver and what the business wanted (and the fact that the latter had changed).
Method	No benefits tracking	The fundamental benefit for the company was having this facility operating at or near capacity, making innovative new drugs and supporting the launch and sustainable supply so that they could be a leader within the biopharma industry. This was never articulated within the project nor tracked upon project completion. Even after the facility was sold, the project remained an example of successful project management.

(Continued)

Table 6-4 (Continued)

Failure area	Root cause	Assessment
Supply	Scope defined from the **initial** understanding of business needs	The scope was delivered exactly as defined within the funding request. Key technical decisions made at this very early stage were not challenged later on. The importance of the technical decision to go for a facility which could not handle solvent processing was fundamental to the delivery of future business benefits and should have been identified as such.
Business	No real 'live' link to the business drivers	The Project Manager was not focussed on getting the first successful process/product in the facility – merely handing over a functional entity. Neither he nor his team were aware of the business drivers, nor the factors which would determine beneficial use of the facility.
	No integration of changes in external environment	No one in the team or the business identified the importance of the changing technical environment. The fact that many manufacturing processes would more than likely require solvents, and that other competitor manufacturers were building smaller multipurpose (solvent and aqueous) biopharma facilities was also overlooked.

Learning points

This project illustrates some important aspects of project delivery: active links between the project and the business and the need for an effective sponsor–Project Manager relationship. Taking each in turn:

Active links between the project and the business

Very rarely is the business environment the same at the end of the project as it was at the start. To allow a project to behave as an island is not business sensible, especially with the increasing rate of change in technology and the competitive environment.

The development of a project business plan such as outlined in *Real Project Planning* (Melton, 2008) would have defined the key links between the project and the business. This plan could then have been tracked through use of the Business Plan Review Checklist (Table 3-4). Such a plan would have required the Project Manager to specify the business benefits and understand the link with the project scope.

An effective sponsor–Project Manager relationship

The main link between a project and the business is via the sponsor–Project Manager relationship. To be operating in a new business environment and **not** have an ongoing discussion on the key business issues and decision points as the project progressed seems naive. A key role of the sponsor is to maintain the business 'radar' as far as the project is concerned.

Using a tool such as the Sponsor Contracting Tool from *Real Project Planning* (Melton, 2008) would have defined the relationship, how it would operate and key discussion areas. They could also have used the Roadmap Decision Matrix, also from *Real Project Planning* (Melton, 2008), to define and manage the technical decisions which increased or decreased future facility flexibility. In addition, the Communications Tracker (Table 3-6) and the Business Plan Review Checklist (Table 3-4) would have allowed assessment of whether the appropriate communication was being supported by the sponsor and whether he was truly engaged in the project.

Case study E – petrochemicals facility project

Sometimes projects can fail to deliver on their key measures (usually cost and schedule) but manage to deliver overall benefit to the business. Case study E is an example of such a project.

Situation

The project was for a major petrochemicals complex in the Middle East, manufacturing chemicals widely used in plastic bottles. Market demand in the region was expected to increase in the future. The plant was located adjacent to a refinery and associated feedstock plant which would provide the raw materials. The project was a greenfield site and involved the complete Engineering, Procurement and Construction (EPC) phases.

The company appointed a single engineering contractor to act as the main design company for the initial scoping and feasibility phase of the project. The final approved design consisted of three components: utilities, Plant A and Plant B.

Following the initial design phase, the company decided to select the original engineering contractor for certain areas and go out for tender on others. The tendering process was essentially driven by lowest cost against a very comprehensive set of specifications. The final project structure consisted of three engineering contractors each working to fixed price contracts (Figure 6-6). One acted as the supervising Project Management Contractor, and all won their respective contracts by delivering lowest bid. The customer worked very closely with each contractor and adopted an extremely contractual stance.

Project delivery was characterized by change orders, confrontation and the management of contract variations. The customer maintained that contract variations were not valid as the scope had not changed, whilst the various the engineering contractors attempted to recover costs and margin whenever a change was identified.

A consequence of this level of confrontation was that many common components (such as the automation and electrical systems) became specific to the individual subprojects, rather than standard across the entire project, and that any attempts at standardization were driven down to the subsupplier

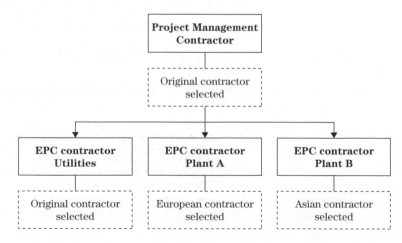

Figure 6-6 Petrochemicals facility project – contractor selection structure

level. Uncertainty over standards and approach, together with regular customer input, resulted in significant rework of designs and delay to the schedule. The overall schedule impact was 9 months in a 2-year programme.

Nevertheless, by the time the facility was brought online, regional demand for the chemical had exceeded expectations and the plant was brought up to full capacity ahead of original market plans. The business objectives had ultimately been met in full, though the project was a failure in terms of schedule and cost adherence.

Root cause

The principal causes of this project failure are shown in Figure 6-7, and Table 6-5 contains more detail of the root causes for each of these.

Learning points

This project illustrates two important aspects of delivery; ensuring contract strategy matches the project delivery goals and the approach to stakeholder management. Taking each in turn:

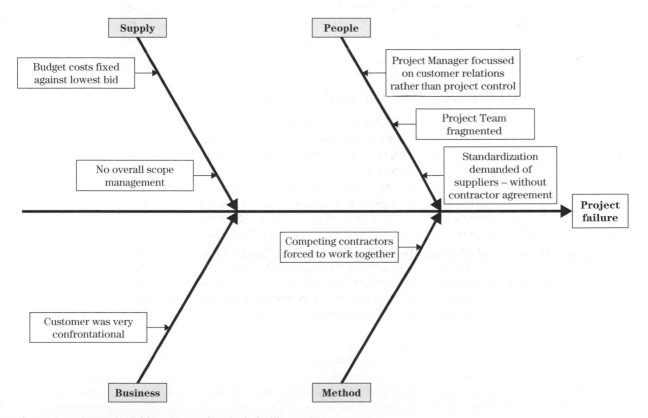

Figure 6-7 Diagnosis Fishbone petrochemicals facility project

Table 6-5 Assessment of failure – case study E (petrochemicals facility project)

Failure area	Root cause	Assessment
People	Project management focussed on customer relations rather than project control	The customer was very confrontational and required significant management time. This meant that the Project Manager did not devote sufficient effort to controlling the overall project in terms of cost, schedule and change.
	Project Team fragmented	The intent of the project was the delivery of an integrated facility. Because of the contractual approach taken, it was impossible for Project Teams to work together to achieve common goals, resulting in duplication of effort.
	Standardization demanded of suppliers – without contractor agreement	The sub-suppliers of each of the three main contractors had the (inappropriate) responsibility of delivering standard solutions across three project components, even though individual components were specified differently. The consequences were rework, increased cost and schedule slippage.
Method	Competing contractors forced to work together	The contractual approach adopted meant that there was little possibility of partnerships between the contracting parties. The customer attempted to act as the arbiter between companies, but this tended to result in requests for variation from one party to another. The overall contract plan was flawed.
Supply	Budget costs fixed against lowest bid	This approach drove much of the project change, as the detailed design generated variations which were not accounted for in the low cost fixed price bids. The customer attempted to enforce the fixed price rigidly, causing significant confrontation and delays in the project as change was contested at all points.
	No overall scope management	Once the components of the project were contractually divided between the individual contractors, it was almost impossible to manage the overall scope. The customer attempted to do this, but his rigid fixed price approach fragmented the scope.
Business	Customer was very confrontational	The high level of confrontation generated behaviours which attempted to manage this (diverting project management effort) and insulate the Project Teams (causing isolation and scope divergence).

Ensuring contract strategy matches project delivery goals

Most of the problems in this project originated from the customer's insistence on an aggressively managed tendering process to multiple contractors. The consequence was that the integrated project requirement was fragmented into multiple silo projects with little co-ordination between the lowest cost bidders.

Using tools such as the Contract Tracker (Table 5-9), Supplier Performance Evaluation and Tracker (Tables 5-10 and 5-11) and the Project Control Tracker (Table 5-15) would quickly identify problems with supplier delivery issues and their impact on other aspects of project delivery. However the problems in this project started during the business case development stage (the business drivers causing the lowest cost bid approach) which were then compounded in the planning stage (development of an inappropriate contract plan). Using tools such as the Contract Plan and the Supplier

Selection Matrix from *Real Project Planning* (Melton, 2008) would have supported a real challenge in this situation.

Approach to stakeholder management

Although the customer had adopted a fixed price project model, he insisted on a very hands-on approach to the project. This perceived interference absorbed significant project time, with the Project Manager having to manage the customer and was contrary to the contract terms agreed.

If a stakeholder management plan had been developed for the overall set of projects, a steering team would probably have been formed to address the issues of project co-ordination and customer management.

Using a tool such as the Stakeholder Engagement Tracker (Table 3-13) identifies stakeholder issues and focuses the team on developing action plans to deal with them. However once again the problems began in the planning stage (no stakeholder management plan). Using a tool such as the Stakeholder Management Plan from *Project Management Toolkit* (Melton, 2007) would have identified these stakeholder issues much earlier.

And remember . . .

➤ A Project Manager can learn a lot from failure – whether it's your own or not. Ensure that every failure is treated as a positive learning experience.

➤ Be honest – in which quadrant in Figure 6-1 would you put your last project?

7

Case Study One: capital engineering project

This case study has been chosen due to its complexity. The project involved multiple sites across several countries with many contracting parties – each of whom had different views on how the project should be delivered. These factors impacted the delivery of the project and illustrate the complexities of the project delivery process. The case study is largely developed from the 'above-site' point of view.

Situation

The company is a major pharmaceutical manufacturer who was predicting a substantial increase in volumes of a newly launched product. Prior to this project, manufacturing was located at one North American facility which was capacity constrained and therefore a potential risk to supply. The key market for the product was North America, so full Food and Drug Administration (FDA) approval was required for any new facility. However, it was expected that in the years post-launch, demand would also become strong in the European and Asian markets.

A project was proposed to significantly increase capacity for manufacturing the finished product (a variety of tablet dose forms) and to diversify the capacity across a number of production locations. The project that was finally approved was budgeted at €103 million and involved 4 locations – 2 in North America and 2 in continental Europe.

Development of the business case

The company's commercial section was forecasting a significant increase in sales of the product over the next 3 years. This was mainly in the North American market (where the product had been launched), but indications were that the forecast trend would be replicated in other markets.

The company held a high-level workshop on ways in which the increase in capacity could be achieved in what was a relatively tight timescale. At the same time, launch stocks for the product were being produced at a small facility in North America. This facility was capacity constrained, and it was not feasible to increase its capability without introducing a significant threat to its ability to produce launch stock.

The business justification for the project fell into two areas:

1. *Security of supply* – The project would diversify production from one site to a variety of sites, thus reducing the risk that supply would be restricted.
2. *Financial* – By delivering the increased volumes, the payback period for the project was estimated to be less than 3 years.

Development of the business case was complicated by the need for confidentiality and speed. Thus the scope for the project was developed by a small team using industry norms for cost calculations (at ±20% cost accuracy in most cases). Due to the size of the project, board-level approval was necessary to proceed.

Pharmaceutical practice and approach to delivery

In order to manufacture a pharmaceutical product, a company must meet specific requirements which demonstrate that the manufacturing process for the drug is safe and effective for patients. Any facility which manufactures such products requires a specific licence to produce the product for a market. Since a significant proportion of the drug product would be destined for the US market, all of the facilities involved would require inspection and approval by the FDA.

Each new facility in this programme of work would require similar inspection to meet regulatory expectations. There was, therefore, a perceived benefit in making the facilities as similar as possible to minimize the effort required to support inspections. This was at odds with the company's normal approach to project delivery, which was to allow sites to fully own and execute the project. Historically, this has led to a degree of localized preferential engineering and the proliferation of multiple standards for similar projects.

A compromise was agreed: It was not considered necessary (or indeed practical) to demand identical new facilities on each site, particularly given the existing capabilities of each site and its preferred suppliers. Moreover, the regulator would be more concerned with those parts of the process which were considered process critical, than the differences in general infrastructure.

Equipment is considered process critical if it could have an impact on patient safety, and the decision was therefore taken to use the same suppliers for this equipment, with the remaining supporting infrastructure being developed locally.

There was one further technical constraint on the design: each facility had to meet specified containment standards. This was not necessary for the current product, but was considered important for potential future activities.

A technical strategy was created to ensure consistency across sites, and a contingency fund (10%) was maintained above-site. This fund was created to serve three purposes:

1. To provide a contingency to compensate for the likely error in the initial cost estimates (the project could not be estimated using the usual approach and was largely estimated using historical experience of similar projects, industry norms and key factors such as overall floor area).
2. To remove problems if sites attempted to change the technical specifications and use alternatives to the agreed standard.
3. To provide the programme management with funds to smooth the 'bumps' in any of the individual projects.

Project governance

Due to the complexity of the project, a programme based organization structure was developed (Figure 7-1) and incorporated into the Project Delivery Plan (PDP) and comprised three key levels:

1. *Programme Steering Team* – responsible for overall governance of the funding and looking at the market, in terms of demand forecasts for the pharmaceutical product, and regulatory approvals. This team looked externally into the business to ensure that the project continued to be the 'right' project.

2. *Project Delivery Steering Team* – a small subgroup of 3 people led by a senior member of the Programme Steering Team was charged with ensuring a consistent approach to project delivery between the sites. Each geographic set of sites had a senior Project Manager designated to the role of coordinator. This team looked internally into the projects to ensure that they were delivered 'right'.

3. *Site Project Managers* – all sites had their own local Project Manager and Project Team. All sites were required to follow the same project management standards and procedures during delivery. These were integrated into the site project delivery plans and explained how the project would be kept 'in control'.

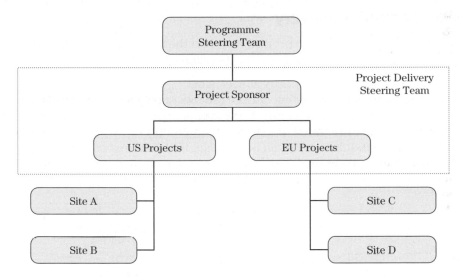

Figure 7-1 Capacity expansion project structure

Project delivery

Because of the size, complexity and number of stakeholders in this project, assessing and mitigating these type of risks was the most significant part of the overall project supervision. Once the project entered detailed design and construction phases, the individual sites would assume the majority of day-to-day control, and the Project Delivery Steering Team was tasked with ensuring that any local problems did not jeopardise overall programme objectives. Therefore, each level of the project organization needed to clearly understand the project benefits required and how these were linked to the Critical Success Factors (CSFs) they would be delivering.

Project CSF and benefits delivery

Figure 7-2 shows the benefits map for the project, based on the main organizational goal of supplying a growing market demand and confirms the key benefits criteria for the project:

- **Capacity** – achieving the required capacity at each site.
- **Supply** – achieving the required lead-time for the product supply at each location.
- **Risk reduction** – minimizing and/or eliminating customer complaints (product quality or delivery issues) and regulatory issues (internal or external audit points).
- **Finance** – achieving the required return on investment, both in terms of starting supply on time with the required capacity and operational cost to manufacture, as well as achieving forecast sales levels.

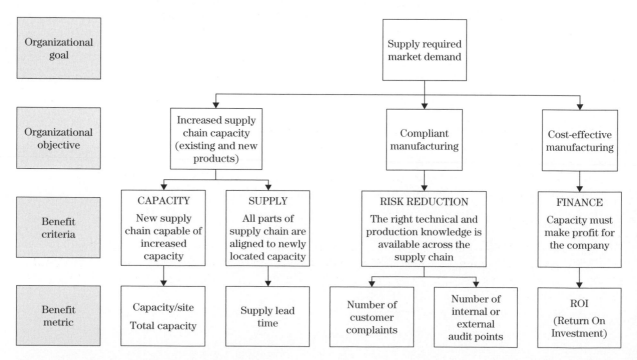

Figure 7-2 Capacity expansion project benefits map

The benefits map enables the identification of the path of success for the project and consequently the Critical Success Factors (CSFs).

Critical success factors (CSFs)

The overall project vision of success was to ensure secure supply to market for the product within cost and time constraints. Five CSFs were identified for the project, and the rationale for their selection is shown in Table 7-1.

Table 7-1 Capacity expansion project CSFs

CSF	Rationale
CSF 1 – Sustainable capacity	The completed facility expansion must be capable of meeting the immediate business needs and continue to do so for its projected lifetime
CSF 2 – Regulatory approval	The facilities must achieve regulatory approval with the FDA to permit the sale of the product into the North American market
CSF 3 – Market supply	The facility must be capable of meeting the projected market demand
CSF 4 – Common platform technology	Key equipment in the facility must adhere to common standards to facilitate commissioning of equipment and regulatory approval.
CSF 5 – CAPEX limit	The project must be delivered within the capital funds agreed with the Board

The CSFs were fully understood by all personnel and used to support project reviews during delivery. Each team member knew that if a CSF was not achieved, then the whole project could not succeed. For example:

- CSF 1 (Sustainable capacity) – Each site had to deliver its part of the overall global capacity requirements in order for the CSF to be achieved. Any individual site failure impacted overall project success.
- CSF 5 (CAPEX limit) – Each site project needed to be controlled within the agreed limits, and the above-site team needed to effectively manage the contingency funds.

Benefits totalizer

The above-site team were particularly concerned with the frequent review of the project's ability to deliver the business benefits, which would be enabled by the successful delivery of the project. Based on the Benefits Map (Figure 7-2), they used a benefits totalizer to provide a high-level status of benefits realization (Table 7-2).

Table 7-2 Capacity expansion project Benefits Totalizer

Delivery Toolkit – Benefits Totalizer				
Project:	Capacity Expansion	**Project Manager:**	Paul Morrison	
Date:	September Year 3	**Sponsor:**	John Drummond	
Customer	**RAG rating**	**Business process**		**RAG rating**
Target capacity in place and supplying market by January Year 4	Green	Successful FDA audit of new facilities without any major observations		Green
Organizational	**RAG rating**	**Financial**		**RAG rating**
Product and process knowledge spread across 4 sites in the manufacturing network	Green	Project ROI of 2.32 years		Green

In the final 4 months of the project, the benefits realization forecast was 'green' for all benefits indicating a high potential that the benefits would be delivered at the pace and level specified in the original business case. This level of information was reported to the Programme Steering Team so that they could assess how this matched with the changing external environment. Fortunately, the market updates for the product remained within the original forecasts.

Project roadmap delivery

The Project Delivery Steering Team tended to use the detailed project delivery roadmap (Figure 7-3) as their method of tracking progress. They used the major stages as review points across all 4 sites (Table 7-3) and in doing so defined key stage gates:

➤ **Approval stage gate** – There was only one formal above-site stage gate – funding approval. At this point, the Programme Steering Team approved the actual project budget for a particular site.

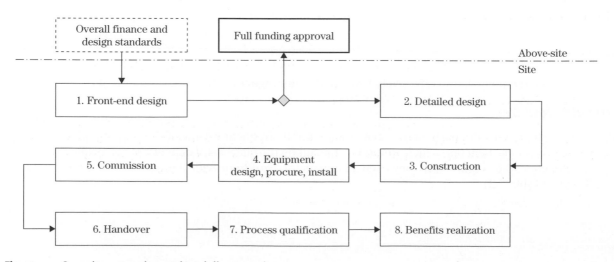

Figure 7-3 Capacity expansion project delivery roadmap

Table 7-3 Capacity expansion delivery stage gate review

Project stage	Review criteria
1. Front-end study	Based on usual company practice, this followed a 2-step approach. Initially the outline design was reviewed for consistency and approach before a more comprehensive design was completed to establish an accurate cost to manage the project against ($+/-10\%$).
2. Detailed design	For this project, this was largely the development of tender packages to the various suppliers. Supplier management processes were crucial to maintaining delivery control.
3. Construction	Activities carried out by an appointed construction contractor in line with established company standards
4. Equipment design, procurement and installation	Specification, procurement and installation of package equipment against the project standards
5. Commission	Integrating the new plant into the existing site infrastructure and verifying installation
6. Handover	At this point, the site operations personnel would assume responsibility for the new facility and prepare for formal testing
7. Process qualification	This is a key stage in the life of a pharmaceutical facility and is the point at which the process design requirements are proven through a series of formal operational tests
8. Benefits realization	At this point, the project is complete and the plant put into beneficial operation and the specified benefits (as detailed in the Benefits Map, Figure 7-2) are checked

⏩ **Review stage gate** – For specified stages the Project Delivery Steering Team reviewed the site projects at the end of one stage and prior to commencement of the next stage. For example, comparison of front-end designs prior to funding approval, cost review at the end of detailed design, construction and commissioning, or technical review at the end of commissioning.

The overall project delivery roadmap has one key input at the start – input of above-site mandated finance and design standards. The company senior management defined the standards and the limits of finance available.

During project delivery, the stages in the roadmap provided site teams with clarity on progress versus plan, as well as allowing comparisons between sites, in terms of both progress and practices.

Project hazard and risk management

During early stages of the project, a detailed risk assessment was not appropriate, so a higher level project hazard analysis was undertaken. This analysis was intended to identify where the major hazards to the achievement of the CSFs would occur and what control mechanisms needed to be in place. A Project Hazard and Critical Control Point (HACCP) analysis was therefore completed. Table 7-4 shows the critical control areas (CCA) identified, and Table 7-5 shows the analysis output in terms of the control mechanisms which were to be used to mitigate risks.

Table 7-4 Capacity Expansion Project HACCP Evaluation

Delivery Toolkit – Project HACCP Evaluation					
Project:	*Facility Expansion*		**Project Manager:**	*Paul Morrison*	
Date:	*June Year 1*		**Page:**	*1 of 1*	
Project vision					
Deliver sustainable capacity across the global manufacturing supply network to enable a sustainable security of supply within the approved project cost and time constraints					
Critical Control Area (CCA) analysis					
Project roadmap step	**CSF1**	**CSF2**	**CSF3**	**CSF4**	**CSF5**
	Sustainable capacity	**Regulatory approval**	**Market supply**	**Common platform technology**	**CAPEX limit**
1. Front end design	*1*			*2*	*3*
2. Detailed design		*4*		*5*	*6*
3. Construction					*7*
4. Equipment design, procurement and installation	*8*	*9*			*10*
5. Commissioning					*11*
6. Handover	*12*				
7. Performance qualification		*13*	*14*		

The 14 critical control areas (CCAs) were considered most important because of the following hazards identified within each project stage:

- **Front-end design** – the inability of the facility to be sustainable (poor design could impact operations), reduced use of common technologies and potential cost escalation.
- **Detailed design** – the failure to gain approval, reduced use of common technologies and potential cost escalation.
- **Construction** – the potential cost escalation.
- **Equipment design, procurement and installation** – the inability of the facility to be sustainable (inadequate equipment), failure to gain regulatory approval and potential cost escalation.
- **Commissioning** – the potential cost escalation.
- **Handover** – the inability of the facility to be sustainable (operations cannot use it).
- **Performance qualification** – the failure to gain regulatory approval and failure to make product to supply the market.

With the CCAs identified, the next step of the HACCP process was to use the Project HACCP Hazard Analysis and Control tool to determine the critical control points (CCPs), the critical control limits

(CCLs) and the hazard identification and control mechanism. Table 7-5 contains partial results of this analysis for project roadmap steps 1 to 7 (Step 1, the front-end design, was completed prior to full project initiation) and also demonstrates the current level of control.

Table 7-5 Capacity expansion Project HACCP Hazard Analysis and Control tool (extract)

Delivery Toolkit – Project HACCP Hazard Analysis and Control							
Project:		*Facility Expansion*		**Project Manager:**		*Paul Morrison*	
Date:		*During stage 4*		**Page**		*1 of 1*	
Project vision							
Deliver sustainable capacity across the global manufacturing supply network to enable a sustainable security of supply within the approved project cost and time constraints							
Hazard	**CCA number**	**CCP (Y/N)**	**Critical Control Limit (CCL)**			**Hazard identification**	**Control mechanism**
			Measure	**Target**	**Actual**		
Supplier design errors	4, 8	Y	Errors per package equipment	2	0	Design review	Supplier management
Poor user specification	4, 8	Y	Design queries from supplier	No major	5 minor	Design review	Project technical standards Supplier management
Performance failure	8, 9, 13, 14	Y	Test failures	5% on test 0% on handover	0	Test process	Supplier management Commissioning management
Site preference for specific equipment	5, 8, 9	Y	Adherence to project standards	No deviations	6 minor	Procurement review	Project technical standards Procurement management
Operations fail to accept plant	12, 14	Y	Handover punch list	No major	0	Operations involvement in key project phases	Stakeholder management
Commissioning overruns	11, 14	Y	Commissioning failures	No major	0	Commissioning process	Commissioning management

This analysis was undertaken at a high (programme) level and allowed the identification of hazards to the overall programme, rather than specific local project risks. This is an important distinction within this type of project, since the Project Delivery Steering Team (Figure 7-1) needed to know where to focus its efforts in order to maximize the potential for programme success. During stage 4 of project delivery, the HACCP review demonstrated that the critical control points (CCPs) were under control, thus effectively managing hazards which would have impacted the whole programme. Individual projects used a failure mode effect analysis (FMEA) approach for local risk control and mitigation.

The major hazards to the successful delivery of the project identified in this partial analysis highlight three mechanisms on which to focus in terms of critical control:

➤ Stakeholder management.
➤ Supplier management.
➤ Project technical standards and commissioning management.

The approach was to develop appropriate high-level guidance and control limits around these 3 control points to ensure that hazards could be identified and addressed rapidly. Appropriate hazard identification and control mechanisms controls were put in place and used during delivery with a high degree of success.

Stakeholder management

The project had a considerable number of stakeholders ranging from board-level management (who sanctioned the original project) to operations (who would ultimately deliver the business benefits). An outline stakeholder map (Figure 7-4) illustrates the range of stakeholders involved in the overall project.

The project HACCP had indicated that stakeholder management was a critical factor in achieving success, and the Project Delivery Steering Team developed a Stakeholder Management Plan to cover overall programme delivery (local stakeholder plans were also in place for site-specific project activities). This plan was converted into a Stakeholder Engagement Tracker and used as a key control mechanism during delivery.

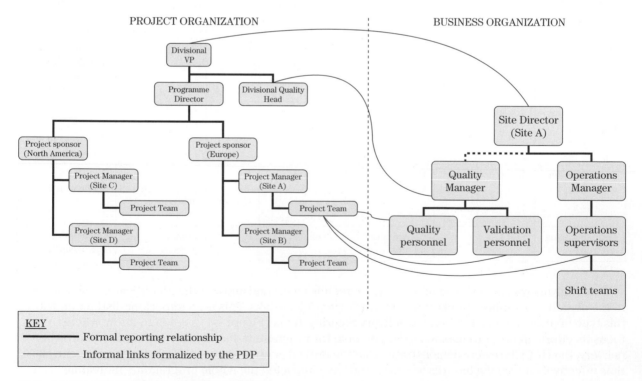

Figure 7-4 Capacity expansion – project stakeholder map

During the early delivery stages the Stakeholder Engagement Tracker (Table 7-6) clearly identified a hazard to the project through the lack of engagement of key operational personnel.

Table 7-6 Capacity expansion project Stakeholder Engagement Tracker

Delivery Toolkit – Stakeholder Engagement Tracker					
Project: Capacity Expansion			**Project Manager:** Paul Morrison		
Date: Early stage in delivery			**Page** 1 of 1		
Stakeholder engagement goal					
Ensure that stakeholders are sufficiently engaged with the project so that it delivers a sustainable, secure supply of product within time and cost constraints. Ensure that stakeholders are managed in a way that delivers the required level of engagement.					
Individual stakeholder engagement analysis					
Stakeholder	**Type**	**Target level**	**Activity**	**Current level**	**Mitigating actions**
Divisional Senior Director	S(a)	High	Weekly briefing on programme status	High	None
Divisional Head of Quality	S(r)	Medium	Circulated with project reports. Occasional attendee at project briefings.	Low	Arrange 1:1 meeting with steering team member. Identify any concerns. Possibly persuade to present impact to next quality forum.
Site A Director	T	High	Sees project reports Heavily involved in other site issues	Low	Arrange 1:1 and project review. Bring him into weekly review with Divisional Director
Site A Project Manager	A	High	Frequent contact with steering team	High	None
Site A Operations Manager	T	Medium	Limited engagement – does not see this as 'his problem' at the moment. Likely to want to wait until the project is ready for handover.	Low	Need to get him engaged into site steering team and support for planning future operating state
Site A Quality Manager	T	Medium	No engagement	Low	Use Divisional Quality Director and quality forum to raise level of engagement
Summary stakeholder engagement status					
Key stakeholders involved project delivery are not adequately engaged. Stakeholders at the operations end are not well engaged and this may have an impact on the handover activities. Unless engagement in both the quality and operations organizations can be increased there is a significant hazard to the success of the project.					

⏺ The engagement of site personnel (quality and operations) was lower than target, and this could have an impact on sustainability of business change.
⏺ In part, this was a consequence of low engagement with the Site Director and the Divisional Quality Director. In essence, if the project is not important to these key stakeholders, it is not important to lower-level personnel.

The mitigating steps developed from this were aimed at building the level of key stakeholder engagement and thereby meeting on-site targets. The ongoing issue with key stakeholder groups required management at above-site and site level:

⏺ *Quality organization* – All levels in the quality organization needed some intervention in order to get appropriate engagement with the project. The root cause for the poor engaging behaviours of the reinforcing sponsor (Divisional Quality Head) was ultimately identified as an ongoing power struggle between him and the operational organization: a perceived conflict in what each part of the organization considered to be an appropriate level of quality.
⏺ *Operational Organization* – Local operational teams at site level did not want to be engaged until the project neared completion and needed to be persuaded of the benefits of earlier involvement. Their attitude was based on previous poor experiences of above-site projects, and consequently they felt that they had little influence on the ultimate outcome.

During delivery the mitigation plans were put into place with the following successful outcomes:

⏺ Once local and above-site teams had understood the reasons for conflict regarding quality standards, and these had been eliminated, the senior stakeholders were better able to engage with the project and demonstrate that engagement within their parts of the organization.
⏺ Local teams were able to get operational representatives involved in the specification of equipment and then gain agreement of their participation in commissioning.

Supplier management

The delivery phase of the project involved the engagement of a number of different suppliers to cover the detailed design, construction, procurement and commissioning steps.

The 'light-touch' approach to project delivery meant that the central group provided steering rather than closely monitored control. Individual sites were, therefore, allowed to establish their own contract strategy based on site preferences and the local market situation:

⏺ One of the sites engaged a single engineering, procurement and construction (EPC) contractor to deliver a complete facility.
⏺ One of the sites adopted a model where the contractor was essentially focussed on the facility construction, whereas the other packages and overall co-ordination and procurement were managed by the local site engineering team.
⏺ The remaining 2 sites adopted a hybrid model with significant EPC engagement, but with the site engineering team providing more detailed input and control to the design.

The first model fitted most closely with the company's project management standards and effectively required less company engagement with the sub-suppliers. The second model required significant involvement from the site personnel in the detail of the project delivery. Both approaches were successful in delivering the overall facility, but two issues became apparent during project execution:

⏺ Issue 1 – design authority and responsibility.
⏺ Issue 2 – supplier performance.

Issue 1 – design authority and responsibility

The buildings required for the new facility needed to meet the requirements of pharmaceutical plant (for environmental conditions), specific levels of cleanliness (to avoid product contamination) and stringent containment specifications to avoid personnel contamination. An important component of achieving these environmental and cleanliness requirements was the use of an appropriate heating, ventilation and cooling (HVAC) systems.

For the site using the full EPC contractor, the overall specification for the building was detailed in the requirements documentation. After an initial design specification was drawn up by the EPC contractor, subcontracts for the HVAC equipment were passed to specialist suppliers. Figure 7-5 illustrates this cascade of design authority.

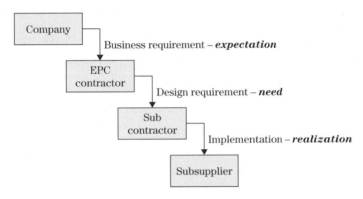

Figure 7-5 Capacity expansion project – design authority cascade

For this class of equipment, customer *expectation* was expressed as an overall performance specification from EPC to subcontractor (the *need*). The subcontractor was then, in effect, delegated design authority for the *realization*. An additional complication was the further delegation of authority for the implementation and design realization to specialist subsuppliers. Typically, this approach can lead to a mismatch between *realization* and *expectation*. Effectively, the design responsibility is delegated along with the authority, even though the subcontracting parties may have insufficient understanding of the company expectations to assume that responsibility.

The project HACCP analysis identified performance failure as one of the key hazards in the project and established the need for supplier management to be used as the control mechanism. The Contract Tracker was utilized to understand how suppliers were performing and to manage project threats. Table 7-7 shows the completed Contract Tracker including the HVAC equipment supply described earlier. Note that it is important within the cascaded design authority model that this contract tracking be established at least against the main subcontractors so that threats and mitigating actions can be correctly established.

Understanding the sub-contract activities and planning an integrated design review allowed the team to ensure that necessary design changes were implemented in a timely manner and that expectations were met.

Table 7-7 Capacity Expansion Project Contract Tracker

	**Delivery Toolkit – Contract Tracker**
Project:		*Capacity Expansion*		**Project Manager:**		*Paul Morrison*
Date:		*February Year 3*		**Page:**		*1 of 1*

Contract	Type	Supplier	Contract tracking			
			Planned status	Actual status	Threat to project	Mitigating action
Granulator – drier system	Direct supplier contract	Supplier A	60%	55%	Low	None – progressing slightly behind schedule – recovery plans in place
Material handling automation system	Direct supplier contract	Supplier B	40%	20%	**High**	Review meeting with supplier required to resolve design differences and put project back on track
Compression machines	Direct supplier contract	Supplier C	45%	45%	Low	None required
Building – ground works, fabrication, tie-ins	EPC main contract	EPC Company	80%	85%	Low	None required
Coaters	Direct supplier contract	Supplier D	25%	28%	Low	None required
Dispensary	EPC sub-contract	Sub-Contractor E	90%	90%	Low	None required
Washer-drier for IBC/drums	Direct supplier contract	Supplier F	90%	100%	Low	Machines awaiting delivery – building not yet complete to receive
Electrical systems	EPC sub-contract	Sub-Contractor G	80%	80%	Low	None required
Fire protection	EPC sub-contract	Sub-Contractor H	75%	75	Low	None required
Mechanical handling equipment	EPC sub-contract	Sub-Contractor I	65%	50%	Medium	Delays due to building access. Meeting scheduled with team to review access plans
Clean area HVAC system	EPC sub-contract	Sub-Contractor J	30%	30%	**High**	Integrated design review with EPC and sub-suppliers to ensure full expectations are being realized
	**Summary contract status**
	Overall contract status is good. Areas of concern around delays in HVAC and automation design are being addressed by design reviews with all parties.

Issue 2 – supplier performance

The site which maintained local control of much of the design and procurement had significant issues with one key supplier's performance. This site was very highly automated within its existing buildings and wished to build on that capability for the new facility. The preferred solution was to provide an integrated material management system linking warehouse, business and manufacturing systems to remove the paper record-keeping necessary for pharmaceutical manufacture. Technology was chosen which was new to the site, as was the supplier. The site was familiar with a project execution model for automation systems which involved 'iterative' design activities with a high degree of interaction between supplier and site personnel. This expectation of interaction was built on relationships with existing suppliers (providing other technologies) and there was a high level of site expectation that the new supplier would work in this way.

During project delivery, there was a serious mismatch between site expectations and supplier performance, and a formal review was conducted (Tables 7-8 and 7-9).

Table 7-8 Capacity expansion project Supplier Performance Evaluation

Delivery Toolkit – Supplier Performance Evaluation				
Project: Capacity Expansion		**Project Manager:** Paul Morrison		
Date: June Year 3		**Page:** 1 of 3		
Supplier: Supplier X		**Contract reference:** AB123		
Required supply				
Software system for automated material handling. Primary point of control for the process				
Supplier performance assessment				
Performance area	**Ranking**			
Project management and control	⇒ Non existent or inadequate project management ⇒ No evidence of formal controls in place	⇒ Project Manager identified ⇒ Simple project plan in place, no evidence of update/use ⇒ Reliance on individuals rather than process ⇒ Haphazard controls	⇒ Project Manager and team identified ⇒ Project plan in place but limited evidence of update/use ⇒ Project Team managed through meetings – no use of tools ⇒ Inconsistent change control	⇒ Project Manager has formal ownership of project and team ⇒ Single point of accountability for decisions ⇒ Formal project management processes followed ⇒ Rigorous change control	⇒ Full and accountable project management process ⇒ Detailed and controlled processes ⇒ Full visibility of progress, issues and changes
Score		3			

(Continued)

Table 7-8 (Continued)

Delivery Toolkit – Supplier Performance Evaluation

Project:	Capacity Expansion	Project Manager:	Paul Morrison
Date:	June Year 3	Page:	2 of 3
Supplier:	Supplier X	Contract reference:	AB123

Supplier performance assessment

Performance area	Ranking				
Communications	» Difficult to contact or obtain a response » Evidence of poor internal communication » Response regularly inadequate	» Regular communication but often incomplete » Response to queries inconsistent » Reactive	» Fairly rapid response to queries » Generally complete responses, but clarification often required	» Effective communication and relationships » Generally pro-active and complete responses » Little clarification required	» Excellent, open relationship » Complete response to queries » Pro-active and anticipates issues
Score		2			
Flexibility	» Inflexible and reliant on contract	» Some willingness to be flexible, but only short-term	» Willing to be flexible around project demands over medium term	» High degree of flexibility around project and contract matters	» Completely open and flexible – joint partnering arrangement focussed on project
Score		2			
Capability	» Inadequate capability » Consistently missing critical deadlines or milestones » Multiple design or production errors	» Poor capability » Some missing of critical deadlines or milestones » Design or production errors not satisfactory	» Satisfactory capability » Almost no missing of critical milestones or deadlines » Design or production errors not critical	» Good capability » No missing of critical milestones or deadlines » Virtually no design or production errors	» Excellent capability » No missing of any project milestones or deadlines » No design or production errors
Score		2			

(Continued)

Table 7-8 (Continued)

Delivery Toolkit – Supplier Performance Evaluation				
Proje ct: *Capacity Expansion*		**Project Manager:**		*Paul Morrison*
Date: *June Year 3*		**Page:**		*3 of 3*
Supplier: *Supplier X*		**Contract reference:**		*AB123*
Supplier performance assessment				
Performance area	**Ranking**			
Delivery	⇒ Frequently capacity constrained resulting in significant schedule problems ⇒ Expediting regularly required	⇒ Some capacity constraints with some impact on schedule ⇒ Some expediting required	⇒ Generally unconstrained and able to meet schedule ⇒ Limited expediting required	⇒ Regular deliveries on schedule ⇒ Limited capacity to reschedule to meet project changes ⇒ Little or no expediting required	⇒ Established track record of deliveries ⇒ Capacity to reschedule to meet project changes ⇒ No expediting required
Score		2			

Table 7-9 Capacity expansion project – Supplier Performance Tracker

Delivery Toolkit – Supplier Performance Tracker			
Project:	*Capacity Expansion*	**Project Manager:**	*Paul Morrison*
Date:	*June Year 3*	**Page:**	*1 of 1*
Supplier:	*Supplier X*	**Contract Ref:**	*AB123*
Supplier profile			

Project management and control
Delivery
Communication
- - - ◆ - - - Actual
—■— Ideal
Capability
Flexibility

Supplier action plan

➤ Project team – acceptable – no further action
➤ Project management and control-acceptable – no further action
➤ Communication – not acceptable – review revised ways of working with supplier
➤ Flexibility– not acceptable – review revised ways of working with supplier
➤ Capability – not acceptable – review root causes and agree an action plan to eliminate
➤ Delivery – not acceptable – review root causes and agree an action plan to eliminate

Schedule a supplier review meeting ASAP.

The team held a combined review session with the vendor using the '5 Whys' approach to establish the root cause of the problems (Figure 7-6).

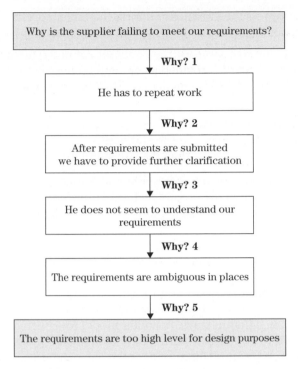

Figure 7-6 Capacity expansion project – supplier '5 Whys' root cause analysis

The review highlighted two key factors in the project:

- The supplier–site relationship was poor as the site continually blamed the supplier for failure to perform.
- The requirements were structured for an 'iterative' design approach where the supplier would work to provide the necessary design detail for implementation. However, the contract was fixed price for design and implementation, and the supplier was expecting a fixed requirement to work against.

The project HACCP analysis had identified poor user specification as a hazard in the project. Unfortunately in this case, the site did not consider its specification to be inadequate as the approach had been used successfully in the past with other suppliers.

The problem was resolved by breaking the link between design and implementation. This ensured that the supplier could complete an adequate design which could be signed off by all parties before the implementation phase was entered. The above-site contingency fund was used to help support this modified activity.

Project technical standards and commissioning management

One of the CSFs for project delivery was the adoption of common technologies in specific areas. Most of these involved the use of the same packaged equipment from approved suppliers. At the front-end stage, the Project Delivery Steering Team had defined a series of technical standards to which individual sites would adhere.

The following major equipment fell into this category:

- Mixer/Granulator – used to form granules of material for the compression step.

- Tablet press – used to compress the granules and form the tablet cores.
- Coater – used to apply a coating to the tablet (the final production step).

In addition, there were specific standards for the containment of the facility.

Whilst the project technical standards provided a common technology basis for the sites to work to, in practice the sites challenged this almost immediately. The challenge was based around the following arguments:

- Specified equipment was unfamiliar to the site.
- Specified equipment was not as good as the site standard.
- Specified equipment was more expensive than the site standard.

These arguments essentially come down to cost: either additional procurement cost or additional training cost. To facilitate adherence to the project technical standards, the Project Delivery Steering Team retained a 10% above-site contingency fund intended to smooth potential cost issues and diffuse site arguments.

The fund was administered by the Project Delivery Steering Team with sites requesting access to additional funds to meet the technical standards. Access to the contingency fund required a detailed business case and justification.

This approach to technical standards followed through to the process of managing commissioning. It was not considered realistic to select standard equipment and then permit sites complete freedom to commission the equipment in different ways. A similar model for specifying the approach to commissioning was developed to avoid this – particularly by mandating the use of an in-house standard commissioning and qualification system. This system defined a standard set of tests for platform equipment which mitigated against local site preferences which may have been either too complex or too simple.

Project reporting and communication

Another key aspect of controlling such a large and complex set of projects was the need for consistent and clear communication and reporting of progress and issues during each stage of delivery. Although the sites used the company's project management standards, they each had their own slight variations of filing project information and reporting it.

The project management standards defined a comprehensive reporting methodology involving the production of a long and detailed monthly report (more than 10 pages) covering the usual factors of cost, schedule and risk. Preparing such a detailed report was something of a chore for the site team and left considerable room for questioning the meaning of some of the detail. There is always the opportunity to lose awkward information within the detail of this type of project report (particularly when communicated to senior management) through the use of selective and different reporting depending on the audience.

The Communications Tracker (Table 7-10) was used to evaluate the effectiveness of communication within the overall project as delivery progressed.

Table 7-10 Capacity expansion project – Communications Tracker

Delivery Toolkit – Communications Tracker				
Project: Capacity Expansion			**Project Manager:** Paul Morrison	
Date: Jan Year 1			**Page:** 1 of 1	
Communications goal				
The communication of clear, accurate and appropriate information to each type of stakeholder, so that they remain engaged in the project and so that they are able to fulfill their own role (in support of successful project delivery). Achievement of this goal is via achievement of the stakeholder engagement targets.				
Key message	**Audience**	**Activity**	**Feedback**	**Mitigating actions**
Project progress – cost and schedule	Divisional VP	Weekly summary via telecom	Immediate from the meeting. Some questioning of the level of detail contained in the reports	Reports take considerable effort to tailor for this meeting – need a simpler way to achieve the result. Generate a lean report
Project progress – cost and schedule	Site leadership team	Infrequent attendance at site meetings	Immediate from the meeting. Lack of understanding of the 'big-picture'. Meeting tends to go over old ground and repeat things.	Get standing slot on weekly site team meeting. Take executive level report and use as basis of this update.
Project progress	Site Project Team	Notice board – visual factory	Ad-hoc from team members commenting on performance	Ensure team members attention drawn to visual factory at regular project meetings
Communications summary				
Generally good communication although improved and more consistent engagement with site team would help. Need to ensure that communications are lean and that only information that helps the audience to understand, make a decision or provide support is delivered.				

Lean reporting

It was evident from the Communications Tracker that the effort needed to produce a condensed report for the executive team was difficult to sustain. Furthermore, the CSFs and hazards identified in the project HACCP were cascaded into the steering reports. A consequence of this was that the Project Delivery Steering Team had difficulty in interpreting and filtering the information in the right way at the right time.

As a result, a lean reporting model was established to keep the report short and clear and allow consistent messages to be filtered up the reporting chain and cascaded down into the Project Team. This reporting model used the project benefits map (Figure 7-2) and the project CSFs (Table 7-1) to develop a report based on three key elements:

1. **Business change** – helps link the project to the overall business and focuses on key critical dates which support achievement of benefits realization.
2. **Project hazards** – helps identify the status of risks that may impact the success of the project.
3. **Cost/procurement** – helps clarify the level of spend against the overall CAPEX budget.

The report which was generated (Project Dashboard, Figure 7-7) greatly simplified the communication of status messages at all levels within the organization.

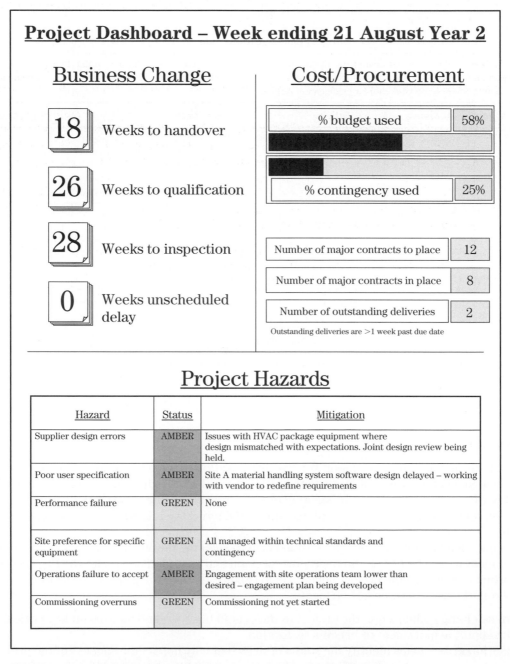

Project Dashboard – Week ending 21 August Year 2

Business Change

18	Weeks to handover
26	Weeks to qualification
28	Weeks to inspection
0	Weeks unscheduled delay

Cost/Procurement

% budget used	58%

% contingency used	25%

Number of major contracts to place	12
Number of major contracts in place	8
Number of outstanding deliveries	2

Outstanding deliveries are >1 week past due date

Project Hazards

Hazard	Status	Mitigation
Supplier design errors	AMBER	Issues with HVAC package equipment where design mismatched with expectations. Joint design review being held.
Poor user specification	AMBER	Site A material handling system software design delayed – working with vendor to redefine requirements
Performance failure	GREEN	None
Site preference for specific equipment	GREEN	All managed within technical standards and contingency
Operations failure to accept	AMBER	Engagement with site operations team lower than desired – engagement plan being developed
Commissioning overruns	GREEN	Commissioning not yet started

Figure 7-7 Capacity expansion project – Lean project reporting dashboard

Other modified reports formed the basis of specific detailed reports within the project and were used as the basis for the Project Dashboard. For example:

- The Supplier Evaluation Tracker was combined with purchase order information to provide a snapshot of procurement.
- The cost performance of the project was combined with the Benefits Totalizer to create a view of project cost and benefit for use with senior management.

Visual factory

Within the Project Team, communication was continuous and based on the specific project stage at any one time:

- During the design phase, the project office at one site had one wall dedicated to tracking design deliverables in terms of their production, review and approval. They also had a notice showing the next critical milestone which the team were working to and used weekly meetings to root cause issues.
- During the construction phase at another site, the construction office used a layout of the facility with various markers to highlight the status in different areas of the facility. This helped to identify who was working where and if the correct permits were in place for this.

The use of the visual factory and lean reporting methodologies meant that team leaders were not spending time producing additional reports. The value-add information which needed to be communicated was instantly generated and available for use.

Conclusions

Overall, project delivery was a success, with all the site expansions being delivered well within the original budget. The use of an above-site contingency fund to smooth the issues with site technical preferences undoubtedly helped minimize the potential for conflict within the project.

The complexity and issues associated with a major pharmaceutical industry project and the need for central co-ordination did mean that overall project delivery was slow – just over 3 years from approval to qualification. It was difficult at times to maintain focus on the rationale for doing the project in the first place, however the use of the project vision of success and the CSFs supported a clarity of rationale over such an extended point.

Lessons learnt

- Contract strategy impacts not only money but the basis of the design phase – who takes responsibility for the design and where is the risk? The contract style selected needs to match with delivery approach required.

- New technology can cause a mismatch between requirements and expectations. Spend time up-front in technical risk areas, getting agreement not only on the requirements but also on the design and implementation approach.

- Reporting needs to be sized appropriately for the audience. Large complex reports drive inappropriate reporting behaviours and obscure information.

- Delivery control mechanisms need to be customized for a project and linked to the specific hazards identified for that project.

8

Case Study Two: accounts system improvement

When an organization decides that it must improve a particular area of its business operations, it can do so in a number of ways: continuously on the job, incremental improvements or step change improvements. This case study is an example of the latter. In this example, the organization considers the improvement as a project, and treated it as such. This combination of good project management practice and good business process improvement shows how the two can be successfully integrated to maximum positive business impact.

Situation

A medium sized IT consultancy working across the UK and Europe identified some issues with its accounting business process:

- Reports lagged behind the true financial situation – so the management team had developed their own ad hoc systems and spreadsheets to assess their departmental profit and loss status.
- The system was thought to contain errors – missing records, incorrect filing, incorrect data entry and issues with accounts reconciliation.

The company accounts system was administered by the Business Support Manager (Kate Robertson). A key part of her role was compliance with external regulatory needs such as quarterly reporting and taxation requirements. This drove the overall timeline for the accounts system business process. Data entry was therefore left for long periods of time and although there were systems in place for collection of information, these were ad hoc processes that had been put into place over time. There was a feeling that errors were being made within the system, mainly due to missing information or associated documentation. Because of the long waiting time for data entry, it was difficult for senior management to get an accurate picture of the company's exact financial position. Kate wanted the system to be improved as she had to deal with the complaints and issues arising from the process. She recognized that she needed to support the delivery of the project and was keen to provide whatever support she could, whilst also continuing to do her 'day job'.

Project initiation and planning

A Project Team was pulled together from the consultants group, with Fred Jones taking on the role of Project Manager. Initially he found it difficult working with the Managing Director (Mary Black) especially trying identify her requirements. Although she wanted the system improved, she seemed distant from the project, as though she believed that the Project Team should know what they needed to do without her input. Fred enlisted the support of Kate (project champion) and together they worked with Mary until she was better able to understand her role (as the customer) and articulate the problem, as follows:

'The current accounts management and business management system is difficult and time consuming to administer, requiring expert input to work. End of year accounts (produced by the accountant) are often impossible to reconcile with company records. A lack of confidence in our own data means that we cannot check the amounts of tax and VAT paid and must rely on accountant's figures. Accounts management reports are often unavailable when required, either because data has not been entered (due to batching) or the system requires minor fixes because of errors. Gaining financial control is a key organizational goal. This means we need to be fully compliant with regulatory requirements for both operations and record keeping, and our finances need to be both accurate and up to date.'

Project approach

The Accounts System Improvement Project had to address these issues and improve the system appropriately. Project stakeholders needed to be identified and engaged to ensure a full understanding of the existing system, its problems and likely direct and indirect benefits from the improvements. The Project Manager confirmed that the improvement process used would be 'lean six sigma', a process he had used many times for the company's clients. This process involves the robust collection and analysis of data to identify the root cause of issues which can then be appropriately solved. The planned delivery approach used the DMAIC framework (Table 8-1).

Table 8-1 The DMAIC framework

Phase	Phase description	Phase aim
D	Define	Define the problem statement
M	Measure	Measure the current performance, collect data around the problem area
A	Analyze	Analyze data to understand the true scale of the problem and to identify the root causes
I	Improve	Solve the root cause problems by waste elimination or process redesign and then implement the design
C	Control	Measure the situation to see that the change is being sustained

In order to compile the formal business case, the Project Team had already commenced data collection and analysis and confirmed the project rationale including the cost/benefit summary (Figure 8-1).

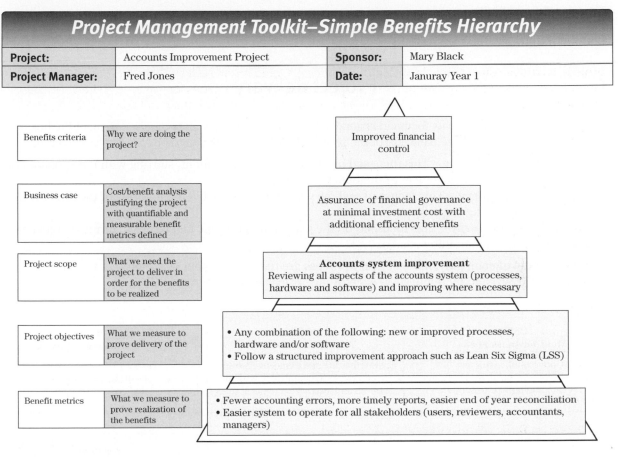

Project Management Toolkit–Simple Benefits Hierarchy

Project:	Accounts Improvement Project	Sponsor:	Mary Black
Project Manager:	Fred Jones	Date:	Januray Year 1

Benefits criteria	Why we are doing the project?
Business case	Cost/benefit analysis justifying the project with quantifiable and measurable benefit metrics defined
Project scope	What we need the project to deliver in order for the benefits to be realized
Project objectives	What we measure to prove delivery of the project
Benefit metrics	What we measure to prove realization of the benefits

Improved financial control

Assurance of financial governance at minimal investment cost with additional efficiency benefits

Accounts system improvement
Reviewing all aspects of the accounts system (processes, hardware and software) and improving where necessary

- Any combination of the following: new or improved processes, hardware and/or software
- Follow a structured improvement approach such as Lean Six Sigma (LSS)

- Fewer accounting errors, more timely reports, easier end of year reconciliation
- Easier system to operate for all stakeholders (users, reviewers, accountants, managers)

Figure 8-1 Accounts project Benefit Hierarchy

The business case was also supported by a detailed benefits mapping session, which confirmed the key goals of the project in terms of benefit metrics (Figure 8-2). Following this level of definition a baseline for current performance could be established.

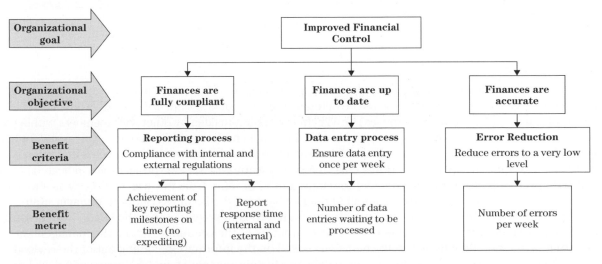

Figure 8-2 Accounts project benefit map

Project delivery

The delivery of the project was focussed around sustainable business change required to meet the business needs and therefore the three elements of the project delivery plan needed to be effectively delivered.

➤ **Business plan delivery** – This focussed on stakeholder management, starting with the sponsor and customer and then management of the business change.
➤ **Set-up plan delivery** – This focussed on how value was to be delivered in terms of the scope and the people delivering the scope.
➤ **Control plan delivery** – This focussed on the management of risk and issues in terms of managing uncertainty and maintaining control.

The key goals of each part of the delivery process are the achievement of the project vision of success (Figure 8-3) and by implication each of the critical success factors (CSFs).

CSF 1 **Meet business needs** Understand customer requirements – Voice of the Customer (VOC) and build into scope	CSF 2 **Appropriate design** Ensure that the design meets customer needs *and* is in compliance with all external and internal accounting rules	CSF 3 **Robust implementation** Delivery of all aspects of the design within time and cost constraints
CSF 4 **Change business ways of working (WoW)** Stakeholder buy-in so that they use the new system and this is checked to ensure sustainability	CSF 5 **Deliver benefits** Define and deliver the changes to specified metrics linked to VOC	**Vision of success** An improved accounts business process capable of sustainably delivering the level of financial control required by the organization

Figure 8-3 Accounts project path of CSFs

Business plan delivery

During the business planning stage, it was determined that Mary, the Managing Director, was the appropriate person to engage as the project sponsor as the outputs of the accounts system are ultimately either direct deliverables to her (for reporting purposes) or will effect decision-making by other staff members within the company. It was therefore considered key to have her full buy-in to the capabilities of the system and for her to have an active role in the implementation of the system both for the data gathering process and roll out. Figure 8-4 shows the overall organizational chart for the company with specific project roles indicated.

Initially, Mary did not understand why the required level of input from her was necessary as she felt that the final user (and therefore customer) of the new system, the Business Support Manager, Kate, would provide this role. It took several meetings with her to get agreement that the key to a successful project was to meet the requirements of the true customers of the accounts system, the management team, that is, the people who will make decisions based on the information coming out of the revised system. In effect the project should be able to deliver a further unstated benefit of improved decision making through the management team having more accurate and current information.

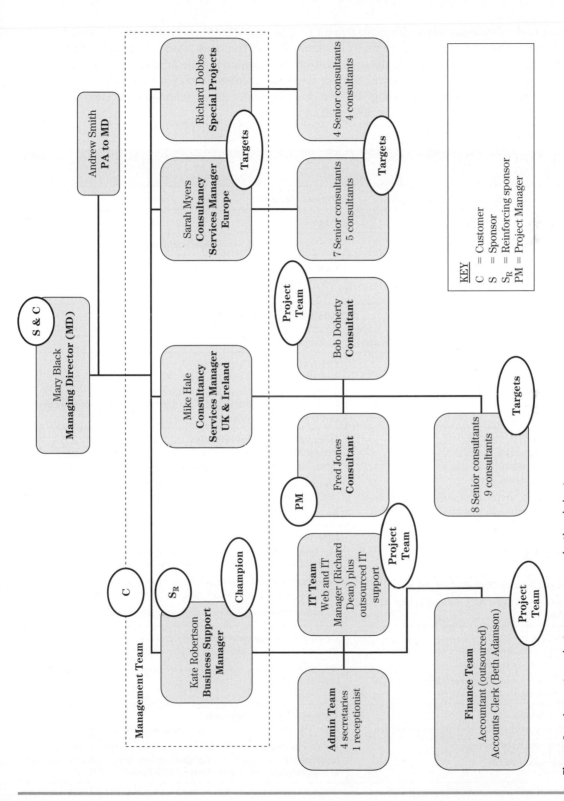

Figure 8-4 Accounts project company organizational chart

The project had to address the problem statement as articulated by Mary and improve the system appropriately. The management team originally assumed that all that was needed was a new IT system – they did not consider the way people in the organization were impacting the accounts business process or how important that business process was to their decision making. The preliminary Project Charter (Table 8-2) was developed in part to manage the senior stakeholder group who were all 'targets' (and ultimately customers) of the change being delivered by the project.

Table 8-2 Accounts project – Project Charter

Planning Toolkit – **Project Charter**	
Project: Accounts Improvement Project	**Project Manager:** Fred Jones
Date: 8 February Year 1	**Page:** 1 of 1
Project description	**Project delivery**
Sponsor Mary Black (Managing Director)	**Project Team** Bob Doherty (Lean Six Sigma Black Belt) Richard Dean (IT Manager)
Customer Kate Robertson (Business Support Manager) – who is also the reinforcing sponsor with specific delegated authority	**Additional resources** Beth Adamson (Accounts Clerk) – with links to the Accountant (outsourced)
Project aim Provide a revised accounts system to meet organizational goal of improved financial control	**Critical Success Factors** As defined in the Path of Success: 1 – meet business needs, 2 – appropriate design, 3 – robust implementation, 4 – change business ways of working (WoW), 5 – deliver benefits
Project objectives Develop a revised company accounts system and business process which meets customer and operational requirements. Develop a monitoring and control system that ensures the revised accounts system is operating as intended.	**Organizational dependencies** Consultant behaviours must be changed in order to ensure information flows into the system in a timely manner. The management team interact with the account process as a part of their report collation – this needs to be considered. Review of the current IT system is a part of this project.
Benefits Fewer accounting errors. More timely reports. Easier end of year reconciliation of accounts. Plus other metrics yet to be fully defined (a part of the design process).	**Risk profile** The time required for data collection to analyze current state at the start of the project may impact the overall timeline. Consultant behaviours may be difficult to change – they always think they know best. Ad hoc systems to get information may not go away.
Final deliverable New/revised accounts business process and the hardware/software to operate this. At this early stage it cannot be presumed that this is purely an IT solution.	**Critical milestones versus deliverables** Revised system operational within 5 months. Target performance within 7 months (project schedule yet to be determined).
Interim deliverables Baseline current performance and targets for future. Determine root causes of issues. Design and test new system.	**Project delivery approach** Lean Six Sigma business process improvement approach. Define voice of the customer (VOC), measure current situation, analyze current situation, design and implement new system, measure new system and implement control and monitoring (DMAIC).

Stakeholder tracking during the first month of project delivery was a key project management activity.

The stakeholder analysis conducted during the business planning stage had already highlighted some areas of resistance – both passive and overt:

- The management team consisted of the four managers who reported directly to the Managing Director. Two of these were openly critical of the time and possible money to be spent on the project.
- The various consultants in the organization were oblivious to the impact of their interaction with the accounts system, even though they spent their working lives dealing with similar issues in other companies.
- The IT Manager (Richard) assumed that the solution would be a new accounts software package and was very sceptical of the Lean Six Sigma approach proposed.

These issues were tracked and after the first month some additional steps needed to be taken (Table 8-3) due to the level of resistance seen:

Table 8-3 Accounts project Stakeholder Engagement Tracker

Delivery Toolkit – Stakeholder Engagement Tracker

Project:	Accounts Improvement Project	Project Manager:	Fred Jones
Date:	10 March Year 1	Page:	1 of 2

Stakeholder engagement goal

Assess current level of engagement as a percentage using the following measures dependent on the role of the particular stakeholder.
- Attendance at project-related meetings – 100% attendance on time and involved
- Access to project-related data – 100% support in obtaining data
- Action completion – 100% completion of requested actions (communication, data) on time and in full

Individual stakeholder engagement analysis

Stakeholder	Type	Target level	Activity	Current level	Mitigating actions
Mary Black – Managing Director	S	100%	Initially she kept cancelling meetings with Fred Jones but recently things have improved. She is starting to provide the necessary design data.	70%	Continue with 1–1's as they are starting to work. Need to be clearer on how/what Mary communicates with her management team to get them to do what is needed.
Kate Robertson – Business Support Manager	S_R & Champion	100%	She is very supportive of the project and is keen to take on new responsibilities that may come from it. Has helped to improve MD engagement.	80%	Fred Jones to get an agreed RACI regarding handover and management of the business changes. Need to be clearer on how Kate manages IT Manager.
Beth Adamson – Accounts Clerk	Target	95%	She is collecting data for the project and understands that she will ultimately operate the revised system	95%	Kate Robertson to ensure that Beth can be made available to support the design phase

(Continued)

Table 8-3 (Continued)

Delivery Toolkit – Stakeholder Engagement Tracker						
Project:	Accounts Improvement Project			**Project Manager:**		Fred Jones
Date:	10 March Year 1			**Page:**		2 of 2
Individual stakeholder engagement analysis						
Stakeholder	**Type**	**Target level**	**Activity**		**Current level**	**Mitigating actions**
Richard Dean – IT Manager	Target	95%	He has not integrated into the small part-time Project Team and is working on his own sub-project		20%	Fred Jones to hold a formal team launch session around the project scope using Bob Doherty's skills in lean Six Sigma
Mike Hale – Consultancy Services Manager	Target	100%	Mike has released two of his team to support the project part-time. He has always been frustrated by the inability of the system to provide timely reports to support his management decisions.		100%	Mike is the most senior manager and need to use his good relationships with his team and fellow managers to get others interested in this project and the benefits
Consultancy Managers (Sarah and Richard)	Target	100%	Neither has reinforced the need for their teams to be available to the project or see it as important		30%	Fred to hold a senior stakeholder meeting and use Mike to reinforce the benefits. Develop a more detailed Benefits Specification Table to support this.
Consultants – all other company employees	Target	95%	Consultants are not interested as long as they get expenses paid. Their managers are not helping them to understand the importance of this project.		20%	They need to provide the required data so that the design is based on the best way for accounts information to flow though the company. Get managers to communicate this need.
Summary stakeholder engagement status						
Getting all stakeholders closer to their target engagement level is critical if the business changes are going to be sustained. Apart from the specific mitigating actions above, a sustainability FMEA should be conducted during the design phase to check that success and sustainability is being integrated into the design.						

- Two members of the management team (Sarah and Richard) were refusing access to their teams in order to collect data (via observation and also interviews).
- The consultant's interaction with the accounts system seemed to be getting worse during the early delivery phase: expense claims were submitted even later than usual, receipts were missing with no explanation and travel booked through company credit cards was being done without the paperwork backup.
- The IT Manager was building a business case for an accountancy software package which the accountant had recommended.

It was clear that more communication was needed to get the key messages across:

- The current system has problems which impact us all.
- Solving these problems will bring benefits to us all.

The management team needed to better understand the benefits and so following further data collection, a meeting was held at which the benefit metrics were outlined (Table 8-4).

Table 8-4 Accounts project Benefits Specification Table

Project Management Toolkit – Benefits Specification Table					
Project: Accounts Improvement Project			**Project Manager:** Fred Jones		
Date: 17 March Year 1			**Page:** 1 of 1		
Potential benefit	**Benefit metric**	**Benefit metric baseline**	**Accountability**	**Benefit metric target**	**Area of activity**
Reporting process	Achieve external reporting milestones	Yes With expediting	Kate Robertson	Yes But no expediting	Design of a new accounts system business process and all associated hardware and software
	Internal reporting lag	4–12 weeks		1 week	
Data entry process	Number of documents awaiting processing	61 (This equates to a wait of up to 2 months)	Kate Robertson	Maximum of nine records	
Error reduction process	Number of errors in the accounts per quarter	26.75	Kate Robertson	4.5	
	Defects per million opportunities (DPMO) Long-term measure	111,581		18,770	
	Sigma rating Long-term measure	2.7		3.6	

In addition, the Project Team conducted an early Sustainability Failure Mode Effect Analysis (FMEA) to understand how to design some of the business changes envisaged by the early design work (Table 8-5).

Table 8-5 Accounts project Sustainability FMEA

Delivery Toolkit – Sustainability FMEA						
Project:	Accounts Improvement Project			**Project Manager:**		Fred Jones
Date:	17 March Year 1			**Page:**		1 of 1
Business change area	**Probability of not**	**Impact of not**	**Ability to detect**	**Sustainability threat**	**Risk priority**	**Mitigation plan**
The way accounting data is issued to the accounts department	4 – Depends on consultants	5	3 – Need to make it more visible	4 – Depends on consultants	240	Review visual factory techniques and how everyone can accept responsibility for meeting benefit targets
The way the accounts clerk deals with the documents	1 – Procedures are generally followed	5	4 – Need to make it more visible	2 – Procedures are generally followed	40	
The ways reports are generated	2	5	3	2	60	
The ways reports are used	3 – Depends on managers	4	5 – Decision making not visible	4 – Depends on Managers and MD	240	Possibly outside of scope but linked to organizational goal-check
Scoring system						
Probability 1 = low 5 = high		**Impact** 1 = low 5 = high		**Detection** 1 = early 5 = late		**Threat** 1 = low 5 = high

What was clear from this early work with stakeholders and the associated business change review was that this project was more than just an accounts issue. It impacted the suppliers of the individual pieces of accounting data (the managers, consultants and suppliers to the organization) and the customers of the collated accounting data (Managers and Managing Director). This is summarized by a SIPOC diagram (Figure 8-5) which looks at:

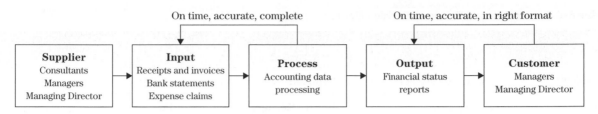

On time, accurate, complete   On time, accurate, in right format

Supplier	Input	Process	Output	Customer
Consultants	Receipts and invoices	Accounting data	Financial status	Managers
Managers	Bank statements	processing	reports	Managing Director
Managing Director	Expense claims			

Figure 8-5 Accounts project SIPOC

S = suppliers of the inputs to a process.
I = the inputs to a process.
P = the process under review.
O = the outputs from a process.
C = the customers of the outputs from the process.

This was another useful communication tool in explaining why the project had a wider impact than might have been expected of an accounting project. What was interesting at this stage of project delivery was the additional benefit criteria identified:

➤ Additional benefit criteria – analysis of financial data so that the management team could make the right decisions for the organization.
➤ Implications of new benefit criteria – the potential impact was wider than the original project and was in effect 'out of scope' but under review by the management team as a part of their business as usual (BAU) operations.

During the remainder of the project delivery (February to June), the tools in the business plan continued to be used to track the status of the benefits and business changes as the design of the latter progressed.

Set-up plan delivery

Recognizing the wider implications of the project, the Project Manager developed a customized roadmap which followed with the DMAIC process. This mirrored in-house company procedures used on client projects and incorporated the following approval stages:

1. Agreement that the initial project idea can provide a real benefit to the company, usually following stakeholder meetings to determine that there is a real need for the project.
2. Approval to proceed with project implementation, considering business case, cost of implementation and other evidence that the project is required.
3. Project close out as determined by sponsor and customer agreement that the project has delivered the required business benefits.

The roadmap had clearly defined stage gates (Table 8-6) which were used during delivery and incorporated additional decisions linked to getting initial buy-in (stage gate 1) and sustainability of the project outcome (stage gate 5).

Table 8-6 Accounts project Roadmap Decision Matrix

Planning Toolkit – Roadmap Decision Matrix				
Project: Accounts Improvement Project			**Sponsor:** Mary Black	
Date: 8th February Year 1			**Project Manager:** Fred Jones	
Stage gate	**Decision**	**Decision by**	**Decision when**	**Data needed**
1	Go or stop project	Managing Director	After project definition	Voice of the customer, agreement that there is an issue to address
2	Go or stop project	Management Team	After analysis of as-is data	Current state; error data; DPMO, sigma rating and control charts. Target DPMO and sigma rating.
3	Go or redesign	Management Team	After design complete	Design proposal
4	Handover or not	Administration Manager	After project delivery	Post-project measurements in place
5	Project complete	Management Team	After benefits delivery	Post-project DPMO and sigma rating, long-term control charting of error rates and batch size in place

Stage gates 1 and 2 were completed during February and early March with the majority of the project time being spent in the design of the accounts system. At this stage, Fred, the Project Manager recognized that he needed to clearly allocate the design tasks to ensure that they were being completed by the team member with the appropriate skills. This in part helped to integrate Richard Dean into the project and to understand that this wasn't necessarily an IT project.

Stage 2 work had progressed according to plan by tracking specific parts of the process and collecting specified data:

➤ Accounts in – tray size (this is a queue of financial data waiting to be processed).
➤ System errors (tracked for 3 months).
➤ Reporting lag (the time taken to respond to a request for current financial information – whether an internal or external request).

This data collection was used to support definition of the business case and the associated benefits specification.

During stage 3 the team had to understand the root causes of the problems and then solve them. During this phase the team comprised:

➤ Core team members – those who had responsibilities during all the project delivery phases (stages 2–5):
  ▷ Fred Jones – Project Manager
  ▷ Bob Doherty – Lean Six Sigma capability
  ▷ Richard Dean – IT capability
  ▷ Beth Adamson – Accountancy capability

Support team members – those who were involved in specific stages to support developing a sustainable design (mainly stages 2 and 3).

During this stage, the sponsor also delegated authority to her reinforcing sponsor, Kate, so that the detailed interim design decisions could be made in a timely manner. The ultimate stage gate 3 decision still required Mary's involvement – although this was a joint decision with Kate. In order to gain stakeholder engagement from her peer group, Kate consulted with them throughout this period.

As the data collection and design progressed, Fred conducted periodic reviews of team progress and one example of this is shown in Table 8-7.

Table 8-7 Accounts project Tracking RACI

Delivery Toolkit – Tracking RACI								
Project:	Accounts Improvement Project			**Sponsor:**			Mary Black	
Date:	17 March Year 1			**Project Manager:**			Fred Jones	
Activity tracking								
Activity	**R** (responsible)	**Y/N**	**A** (accountable)	**Y/N**	**C** (consulted)	**Y/N**	**I** (informed)	**Y/N**	**Gap analysis**
Data collection – accounts office	Beth	Yes	Fred	Yes	Bob	Yes	Kate	Yes	No issues, all collected to plan
Data collection – consultants	All in BAU	**No**	Fred	Yes	Bob	**No**	Kate	Yes	Problems getting data from consultants
Scope development	Bob	Yes	Fred	Yes	Beth Kate Richard BAU	Yes Yes No No	Mary	Yes	Need to engage some of BAU in this activity – build a Critical to Quality (CTQ) Tree
Design development	Bob Beth Richard	Yes Yes **No**	Fred	Yes	Kate BAU	**No** **No**	Mary	**No**	Early days yet – need to get a workshop organized
Communications	Fred	Yes	Kate	Yes	Bob Richard Beth	Yes Yes Yes	BAU	Yes	No issues – communications all going to plan
Action plan								
Need to set up a clearer support team from BAU and involve them in two key workshops – scope development and design development. Then they can be used as champions during implementation.								

This particular review highlighted that further consultation and involvement with the support team was needed:

➤ A consultant from each of the departments was asked to become involved in a 'BAU reps' group.
➤ The core and support team were joined together for a launch session where the current process and its impact on the business was reviewed (Figure 8-6).

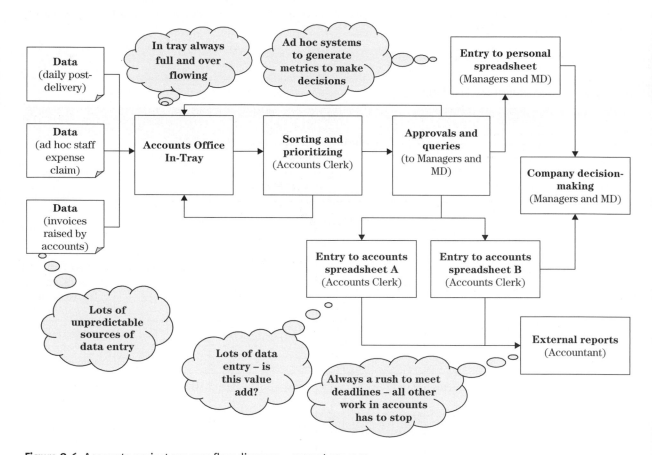

Figure 8-6 Accounts project process flow diagram – current process

The data collected during stage 2 allowed the extended team to see the measures which this current process impacted: errors, lack of timely management information and expediting to meet external targets.

As with any Lean Six Sigma project, it is important that the present situation is fully understood in order to ensure scope design properly addresses current issues and delivers value for the organization. Within this project, scope delivery was supported through the team development of a map of critical features. The core and support team worked together to link the scope of the project to the benefits. They then conducted a root cause analysis on the issue to identify the type of outcome deliverable which would solve the problems in the accounts system (Figure 8-7). The key root causes were:

- Lack of control of the flow of accounting paperwork.
- Lack of understanding of the different needs of the accountant and management.

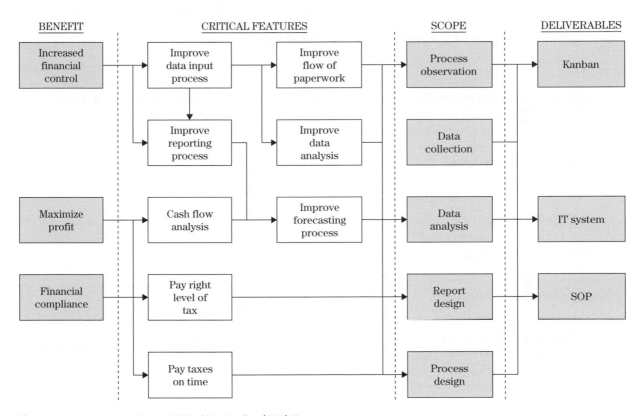

Figure 8-7 Accounts project – Critical To Quality (CTQ) Tree

The level of duplication within the system was seen as a key contributing cause to these issues: At times, a piece of accounting paperwork could return to the in tray up to five times and thus stay unprocessed for weeks if not months.

Design deliverable definition

Based on the identification of the critical features (Figure 8-7), the decision was made to develop a visual factory approach, namely a kanban, with associated Standard Operating Procedure (SOP) and a single spreadsheet as the IT solutions. The extended team were was then able to work together to map out exactly what these deliverables would look like.

Kanban

A kanban is a visual signal for a process to commence or to stop and in this case it will be used to start and stop the accounts data entry process. This kanban could be a series of in-trays or a noticeboard with a fixed number of positions where accounting information (records) can be attached. Each space in the kanban should hold one accounting record and the visual signal needs to be defined based on the volume of data entering the kanban and also the speed at which the data can be processed and therefore leave the kanban. For this project the kanban will be sized to force the accounts process to run once per week (based on data collected at the start of the project).

IT system

A single spreadsheet will be developed so that data entry for each piece of accounting data occurs only once. This spreadsheet will be capable of collating the information for the accountant as well as providing data for the management team and Managing Director. In effect one of the existing accounts spreadsheets will be kept and adapted so that it can provide the required output reports: tax, accountancy and management.

Standard Operating Procedure (SOP)

A new business process (SOP) will be developed to demonstrate how to use the new kanban system to collect and process accounts records using the revised IT system. The SOP will clearly show the roles of all people in the organization.

Design deliverable progress

During the detailed design phase, the Project Manager used a Scope Tracker (Table 8-8) to confirm that the scope continued to meet the needs of the project, and this mitigated against both scope creep and omission. It also records the key decisions regarding the exact format of the final deliverables. For example, based on testing various designs for the kanban it was decided to drop the idea of using in-trays in series and use a noticeboard instead.

Table 8-8 Accounts project Scope Tracker

Delivery Toolkit – Scope Tracker

| Project: | Accounts Improvement Project | | | | Sponsor: | Mary Black | |
| Date: | 20 May Year 1 | | | | Project Manager: | Fred Jones | |

Deliverable	Change?	Scope	Impact?	Critical feature	Impact	Benefit criteria	Impact
Agreed output from the project	*Add, delete, modify*	*Agreed scope to deliver the CTQ feature*	*Impact of change to deliverable on scope*	*A CTQ feature*	*Impact of change on CTQ feature*	*Insert organizational or project benefit*	*Impact of CTQ feature change on benefit*
Kanban	**Modify**	New accounting process	**None** Design has now identified that this is to be a noticeboard not a series of in trays	Improved flow of data entry Increased speed of data entry	**None**	Reduction in errors Reduction in accounting queue	**None**
SOP	No		**None** SOP will be a simple flowchart showing all staff responsibilities	Improved process for data entry, reporting and analysis of reports	**None**	Reduction in errors Improved reporting (accuracy and timeliness)	**None**
IT system	No		**None** IT system will be a bespoke single spreadsheet based on current spreadsheets	Improved data analysis	**None**	Reduction in errors Improved reporting (accuracy and timeliness)	**None**
Management report	**Add**	Supporting management decision-making	**Non-valid scope** The boundary for this project is the generation of a report not the process for analyzing it	Improved data analysis	**None** This change should only go as far as generating a format as an output of the IT system	Improved reporting	Potentially more benefits if scope was extended

Scope change summary

During the design workshop, the idea of the noticeboard kanban was developed. This effectively took away any prioritization and removed all in trays. However, the extension of the IT system to generate a report to support improved management decision-making was felt to be an extension of the project which was outside of scope. This will need to be dealt with if the organizational goal of 'improved financial control' is to be fully met – benefits map needs to be updated to show this.

Control plan delivery

This project was a short fast-track business improvement project which kicked off in February and was handed over in June. Apart from the usual project management tools, the overview tool used by the Project Manager to assess the likelihood of project success was the Project Risk Profile (Table 8-9).

Table 8-9 Accounts Project Risk Profile

Delivery Toolkit – **Project Risk Profile**						
Project: Accounts Improvement Project			**Project Manager:** Fred Jones			
Date: 20 June Year 1			**Page:** 1 of 1			
Risk profile trend						
Risk area	**Milestone review dates**					
	Project start February 8th	**March 17th**	**April 20th**	**May 18th**	**June 1st**	**Handover June 20th**
CSF 1 – Net business needs	Green	Green	Green	Green	Green	Green
CSF 2 – Appropriate design	Green	Amber	Amber	Amber	Green	Green
CSF 3 – Robust implementation	Green	Green	Green	Amber	Green	Green
CSF 4 – Change business WoW	Amber	Red	Amber	Green	Green	Green
CSF 5 – Deliver benefits	Amber	Red	Amber	Amber	Amber	Green
Project risk summary						
At project handover, the forecast is that all CSFs will be achieved. Note that CSF 5 cannot be confirmed until about a month after handover, but signs from the stakeholder review and CSF 4 are that the business will sustain the changes which will ensure full benefits delivery.						

As was indicated during the business plan delivery, during early delivery phases there was an issue with stakeholder engagement in the project outcome. This was clearly seen in the risk profile, which monitored the probability that the project CSFs to be achieved. In addition, the Project Team used a risk table and matrix to monitor risks from their perspective. During the scope development workshop in early April, five risks were highlighted (Table 8-10 and Figure 8-8):

1. **Design issues** – highlighted the risk that the performance required of the new accounts system may not be possible.
2. **Data issues** – highlighted the level of data collection needed during and after project delivery and the risks if this could not be obtained in a timely manner.
3. **Sustainability issues** – highlighted the business change issues inevitable in a project which expects people to operate differently as a result.
4. **Compliance issues** – highlighted the risks when dealing with external regulatory interfaces.
5. **Project delivery issues** – highlighted the consequences of being out of control of both the 'hard' and 'soft' sides of the project.

Table 8-10 Accounts project Risk Table

Project Management Toolkit – Risk Table

Project:	Accounts Improvement Project		Project Manager:	Fred Jones	
Date:	10 March Year 1		Page:	1 of 1	
	Risk description		**Risk assessment**	**Action planning**	
Risk number	**Risk description**	**Risk consequence**	**Occur?**	**Impact?**	**Mitigation plan**	**Contingency plan**

Risk number	Risk description	Risk consequence	Occur?	Impact?	Mitigation plan	Contingency plan
1	Design cannot meet error reduction requirement	(a) Business continues to operate with unacceptable level of accounts errors	Medium	High	Project includes milestone reviews to ensure it remains focussed on objectives – the design phase will involve some mistake proofing	May be able to get additional guidance from external experts if required
		(b) Man-hour costs to fix errors are not reduced	Medium	Medium	Project metrics will be set early on in the project to monitor man-hours	
2	Data collection during and post project will take longer than planned due to the quantity and quality of data required	Project timeline will extend	Medium	High	Data collection will have its own dedicated planning session and link to the way the data was collected to generate the business case	There is already an additional hold point in the project roadmap to prevent it from progressing too far without a robust data baseline
3	Project change may not be sustainable	Old habits will return after start-up and the new system may not get used	Medium	Medium	All stakeholders have bought in to the project requirements and appear committed. Stakeholder management is taking place. Measures have been set up to check sustainability.	Ongoing data will be reviewed periodically, with reinforcement of the required new practices discussed when necessary
4	New system is not compliant with accountancy regulations	Company will not be financially compliant	Low	High	The team has access to the company accountant and the Business Support Manager who will be heavily involved in the project	
5	Uncontrolled project delivery	(a) Don't achieve project objectives	Low	High	Use of an experienced Project Manager who has a strong track record of successful business change	
		(b) Don't involve or engage stakeholders	Medium	High	Have a Stakeholder Management Plan	

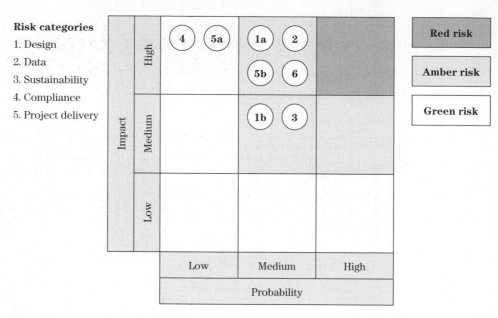

Figure 8-8 Accounts project Risk Matrix

Apart from the risk management tools, the Project Manager used a number of tools and techniques to control project delivery. These were frequently reviewed with either the sponsor and/or the Project Team (Table 8-11).

Table 8-11 Accounts project Control Tracker

<table>
<tr><td colspan="6">Delivery Toolkit – Project Control Tracker</td></tr>
<tr><td>Project:</td><td colspan="2">Accounts Improvement Project</td><td>Project Manager:</td><td colspan="2">Fred Jones</td></tr>
<tr><td>Date:</td><td colspan="2">20 May Year 1</td><td>Page:</td><td colspan="2">1 of 2</td></tr>
<tr><td colspan="6">Cost</td></tr>
<tr><td>Area</td><td>Control tool</td><td colspan="2">Current status</td><td colspan="2">Action/mitigation and forecast status</td></tr>
<tr><td>Spend as planned</td><td>Man-hour cost plan</td><td>Spent 750 man-hours out of the 800 man-hour budget (excludes contingency)</td><td>Amber</td><td>Forecast that a final 100 man-hours will be needed</td><td>Amber</td></tr>
<tr><td>Spend when planned</td><td>Cash-flow profile</td><td>The cumulative man-hour spend profile matches with expectations with the exception of the use of an additional 40 hours during workshop preparation</td><td>Amber</td><td>Forecast profile agrees with the cost plan – minor overspend of less than 10%. Man-hours are not to be capitalized.</td><td>Amber</td></tr>
</table>

(Continued)

Table 8-11 (Continued)

Delivery Toolkit – Project Control Tracker					
Project:	Accounts Improvement Project		**Project Manager:**		Fred Jones
Date:	20 May Year 1		**Page:**		2 of 2
Schedule					
Area	**Control tool**	**Current status**		**Action/mitigation and forecast status**	
Spend risk	Cost run-down (contingency spend)	Contingency of 80 man-hours will be used in the final month	Green	None – 30 man-hours will be available for the closeout/ sustainability	Green
Done when planned	Gantt Chart with critical milestones	All stage gate milestones achieved to date	Green	Forecast handover on June 21 as per plan	Green
Done by who planned	RACI Chart and resource plan	Extended Project Team completing all required actions. Part-time nature of project involvement causing some efficiency issues and minor non-critical delays	Amber	Request additional time from each team member for the final month (already included in man-hour forecast)	Green
Schedule risk	Schedule risk rating	No red risks, 1 amber risk and 4 green risks	Amber	Amber risk relates to ability to get kanban training completed – additional resource above should mitigate	Green
Change					
Area	**Control tool**	**Current status**		**Action/ mitigation and forecast status**	
Stakeholder engagement	Stakeholder Engagement Tracker	Stakeholders are engaged to the appropriate level although there are still small pockets of disinterest within the consultant teams	Amber	Integrate suitable checks into the sustainability checklist to ensure that the behaviour of these individuals is tracked	Green
Deliverable revisions	Deliverable control chart	Kanban – Revision 3 – pilot to be generated to test SOP – Revision 2 – needs to be used in training IT system – Revision 1	Green	Ensure that no further revisions are completed as all need testing during pilot implementation and handover	Green
Scope control change	Scope Tracker	An additional scope area has been proposed, but this does not appear to link overtly to the goals of this project (management decision-making process)	Amber	There is a link between the proposed scope and the organizational goal which this project supports – potentially raise as a new project with dependencies to this project	Amber
Project control summary					
The project remains in control with a forecast outcome of sustainable change with minimal man-hour overspend.					

The project progress measurement used the in-house company systems which the consultants would usually use to control a client assignment. For this project this involved:

- *Cost* – Project man-hours were monitored via the existing company time sheet system.
- *Schedule* – Deliverables progress versus schedule was tracked against a baselined Gantt chart.
- *Cost and Schedule* – Deliverables progress versus forecasted man-hours was tracked. Schedule was developed based on deliverables using company norms for completion of these deliverables and associated activities.

The Project Manager concluded that at handover, the project objectives had been successfully delivered and there was a robust forecast that the benefits would follow. On June 21, the project deliverables were handed over:

- **Kanban** – Effectively a 'hanging in-tray' with slots for each new piece of accounting data to be inserted by the data provider. Once she saw the signal to commence, the Accounts Clerk would simply deal with each slot in sequence until the kanban was empty.
- **SOP** – a flowchart on the wall of the accounts room and all offices. Everyone in the company had been trained in this and their role in its successful use.
- **IT system** – a modified bespoke spreadsheet used only by the Accounts Clerk, which delivered accounting and management reports for use by the Accountant, Management Team and Managing Director, as appropriate.

Conclusions

Following project completion, the benefits and business change behaviours were tracked to ensure that success was sustainable. Post-project data collection focussed on the same metrics as beforehand and concluded that benefits were either meeting target or trending that way.

The Accounts Benefits Tracker (Figure 8-9) was issued to managers and also displayed in the accounts office.

Accounts System Improvement Benefits Tracker		21 June Year 1	
Reporting metric	External reporting achieved with no expediting	Internal reporting lag of 1 week	green
Data entry metric	Average number of documents waiting to be processed is 5. *Note*: the first month has been lower than average in terms of document processing		green
Error reduction metric	Errors per month of 7	Process sigma rating of 3.2	amber

Figure 8-9 Accounts project Benefits Tracker

In addition, a sustainability check was conducted to see if all aspects of the changed business process were being sustained (Table 8-12).

Table 8-12 Accounts project Sustainability Checklist

Project Management Toolkit – Sustainability Checklist					
Project:	Accounts Improvement Project		**Date:**	June 21 Year 1	
Project vision					
The input of data into the accounts system will be controlled on a weekly basis, with a maximum batch size of nine. Error levels will be no more than the agreed reasonable level (Sigma Rating 3.6).					
Sustainability review information					
Previous sustainability review	n/a		**This sustainability review**	June 21 Year 1	
Project representative	Fred Jones		**Customer representative**	Mary (MD) Kate (Business Support Manager)	
Sustainability checks					
Check number	**Check**	**Target (sustained change)**	**Last review**	**This review**	
1	Is data entry kanban being used properly? Is data being placed in it? Are there piles of data next to it?	There should be no piles of data near the kanban, all data should be entered into it	n/a	Kanban is in use – no piles of data	
2	Are personnel adding data to kanban at earliest opportunity as per procedure?	Data should be entered into the kanban as soon as possible	n/a	Three members of staff found to be holding expenses receipts over a month old	
3	Is data kanban being emptied as indicated by amber (start data entry now) and red indicators (must start data entry now)	Kanban should enter amber zone once per week as this triggers data entry process and enter red zone occasionally	n/a	Kanban has entered amber zone four times and red zone once	
Summary comments and next steps					
Kanban is in use, and the only current issue is around personnel holding onto information that should be entered into the kanban. All checks will be monitored further to ensure that the system is being used properly.					
Is the change completely sustained?	No		**Date of next sustainability check**	August Year 1	

This improvement project demonstrated to the management team the importance of identifying the real root causes of an issue and not to make a leap to the most obvious solutions. In this case the obvious solution was a new IT system, when it hardly needed changing at all. The root cause was the way that the accounts data flowed into and out of the accounts process and this was only identified through robust data collection and analysis. This project was clearly a success and delivered all tangible and intangible benefits. Fundamental to this success was the change in behaviour at every level in the organization:

- The consultants submitted expense claims containing the required receipts.
- The Business Support Manager raised invoices, paid expenses and supplier invoices as required.
- The management team approved payments and invoices as required.
- The kanban approach (Figure 8-10) drove input behaviours from the management as well as prioritizing the BAU work for the Business Support Manager.

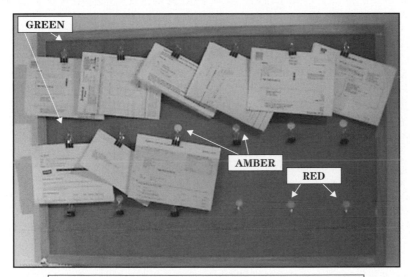

- Green – accounts records do not need to be processed
- Amber – signal to start processing
- Red – MUST start processing

Figure 8-10 Accounts project – the accounting kanban

Lessons learnt

- No matter how small a project, there is value in using appropriate project management tools and techniques to structure delivery, manage uncertainty and deliver benefits. The trick is to choose the 'right' set of tools and techniques.

- Business change is a fundamental part of any project and should be integrated into the 'harder' aspects of project delivery.

- Lean Six Sigma is a powerful business improvement technique, which benefits from a project-structured approach to manage uncertainty and deliver success.

9

Appendices

Appendix 9.1 – The 'Why?' Checklist

Project Management Toolkit – 'Why?' Checklist

Project:	<insert project title>	Project Manager:	<insert name>
Date:	<insert date>	Page:	1 of 1

Sponsorship

Who is the sponsor? (The person who is accountable for the delivery of the business benefits)
<insert the name of the person who is taking this role>
Has the sponsor developed an external communication plan? (How the sponsor will communicate with all stakeholders in the business?)
<insert any comments on how the sponsor has/is communicating with the business>

Business benefits

Has a business case been developed?
<insert comments on the current status of the formally developed business case which supports the project>
Have all benefits been identified? (Why is the project being done?)
<insert comments on the progress of the articulation of the benefits of completing the project>
Who is the customer? (Identify all stakeholders in the business including the customer)
 <insert comments on the completion of the stakeholder analysis>
How will benefits be tracked? (Have they been adequately defined?)
<insert comments on benefits metrics>

Business change

Will the project change the way people do business? (Will people need to work differently?)
<consider if the project will change the way that 'normal business' is conducted>
Is the business ready for the project? (Are training needs identified or other organizational changes needed?)
<consider what else is being done in other parts of the business related to the project>

Scope definition

Has the scope been defined? (What level of feasibility work has been done?)
<insert comments on the accuracy of the scope of the project>
Have the benefit enablers been defined? (Will the project enable the benefits to be delivered when the project is complete?)
<insert comments on how the scope is linked to the business benefits>
Have all alternatives been investigated? (Which may include not needing the project)
<insert comments on all alternatives to this project which have been considered>
Have the project success criteria been defined and prioritized?
<consider the areas of scope which the project requires to be completed in order to deliver the business benefits>

Stage One decision

Should the project be progressed further? (Is the business case robust enough for detailed planning to commence?)
<insert the decision – yes or no – with comments>

Appendix 9.2 – The 'How?' Checklist

Project Management Toolkit – 'How?' Checklist

Project:	*\<insert project title\>*	Project Manager:	*\<insert name\>*
Date:	*\<insert date\>*	Page:	1 of 3

Stage One check

Have there been any changes since Stage One completion? (Development of the business case and project kick-off may be some time apart)
\<insert any changes that may impact the delivery of the project or the associated benefits\>

Sponsorship

Who is the sponsor? (The person who is accountable for the delivery of the business benefits)
\<confirm the name of the person who is taking this role\>
Has the sponsor developed a communication plan?
\<insert a comment on how the sponsor intends to communicate with project stakeholders during the project\>

Benefits management

Has a benefits realization plan been developed?
\<insert any data related to the schedule for delivery of the agreed benefit metrics\>
How will benefits be tracked? (Have they been adequately defined?)
\<insert any additional data which further articulates the specific benefit metrics which align with work completed during Stage One\>

Business change management

How will the business change issues be managed during the implementation of the project? (Are there any specific resources or organizational issues?)
\<insert specific plans for the management of the business change associated with the project\>
Have all project stakeholders been identified? (Review the stakeholder map from Stage One)
\<attach the stakeholder analysis work that has been completed\>
What is the strategy for handover of this project to the business? (Link this to the project objectives)
\<insert specific plans for project handover\>

Scope definition

Has the scope changed since Stage One completion? (Has further conceptual design been completed which may have altered the scope?)
\<insert details of the further work which may have been conducted prior to project kick-off\>
Have the project objectives been defined and prioritized? (What is the project delivering?)
\<attach a copy of the prioritized objectives\>
\<insert an updated list of project CSFs\>

(Continued)

Appendix 9.2 – The 'How?' Checklist (Continued)

Project Management Toolkit – 'How?' Checklist

Project:	*<insert project title>*	Project Manager:	*<insert name>*
Date:	*<insert date>*	Page:	2 of 3

Project type

What type of project is to be delivered? (For example, engineering or business change)
<insert the type of project being delivered – note that this is a major category>
What project stages/stage gates will be used? (Key milestones, for example, funding approval, which might be go/ no go points for the project)
<insert the project road map for the type of project within the organization>

Funding strategy and finance management

Has a funding strategy been defined? (How will the project be funded and when do funds need to be requested?)
<insert the funding request requirements – estimate accuracy, funding timeline, authorization process>
How will finance be managed?
<confirm that no additional reporting outside of the project control strategy is required>

Risk and issue management

Have the CSFs changed since Stage One completion? (As linked to the prioritized project objectives and the critical path through the project risks)
<insert updated critical path of success if available>
Have all project risks been defined and analyzed? (What will stop the achievement of success?)
<comment on any high priority risks>
What mitigation plans are being put into place?
<attach a copy of the high priority mitigation plans>
What contingency plans are being reviewed?
<attach a copy of the high priority contingency plans>
<attach a copy of the Risk Table and Matrix>

Project organization

Who is the Project Manager?
<insert the name of the Project Manager who will be delivering this project in line with the project delivery plan>
Has a project organization for all resources been defined? (Include the Project Team and all key stakeholders)
<insert any comments on the project resource situation – capacity or capability>
<Have roles and responsibilities been defined? Attach the RACI Chart and/or project organization chart>

Contract and supplier management

Has a strategy for use of external suppliers been defined? (The reasons why an external supplier would need to be used for any part of the scope)
<insert a copy of the contract plan>
Is there a process for using an external supplier? (For example selection criteria, contractual arrangements, performance management)
<confirm that procedures to manage supplier selection and performance are in place>

(Continued)

Appendix 9.2 – The 'How?' Checklist (Continued)

Project Management Toolkit – 'How?' Checklist			
Project:	*<insert project title>*	**Project Manager:**	*<insert name>*
Date:	*<insert date>*	**Page:**	2 of 3

Project controls strategy

Is the control strategy defined?
<comment on each of the following:
➡ *Cost control strategy*
➡ *Schedule strategy*
➡ *Change control*
➡ *Action/progress management*
➡ *Reporting*
<What methodologies, tools or processes will be used to ensure control?>

Project review strategy

Is the review strategy defined? (How will performance be managed and monitored – both formal and informal reviews and those within and independent to the team?)
<comment on the plan for reviewing project performance during the delivery of the project>

Stage Two decision

Should the project be progressed further? (Is the project delivery strategy robust enough for project delivery to commence?)
<insert the decision – yes or no – with comments on the robustness of the project delivery plan>

Appendix 9.3 – The 'In Control?' Checklist

Project Management Toolkit – 'In Control?' Checklist

Project:	*<insert project title>*	**Project Manager:**	*<insert name>*
Date:	*<insert date>*	**Page:**	1 of 2

Stage Two check

Have there been any changes since Stage Two completion?
<insert any changes that may impact the delivery of the project or the associated benefits since either Stage Two completion or since a previous Stage Three project review>

Business change management

What is the current status of stakeholder management? (Review the original stakeholder map and discuss)
<insert comments on the status of the stakeholder management plan>
How will the business be expected to operate as a result of the completion of the project?
<comment on the main changes to the business as a result of the project>
Is the business ready for the project?
<insert comments on the development and/or delivery of the plans for the management of the business change associated with the project including any sustainability plans>
What is the strategy for handover of the project to the business?
<insert comments on the development and/or delivery of the plans for the handover of the project>

Scope definition

Has the scope changed since Stage Two completion?
<insert details of any changes to the scope since the project delivery plan was approved or a previous Stage Three project review was conducted>
What is the project progress against Stage Two defined and prioritized project objectives?
<insert a current list of project objectives/CSFs with a comment on their progress versus plan>

Project roadmap

What project stages/stage gates have been completed? (Key milestones, for example, project approved, design complete)
<comment on the internal or external stage gates which have been successfully achieved since Stage Two completion or the previous Stage Three project review>

Risk and issue management

Have all project risks been reviewed regularly during project delivery?
<comment on any high priority risks and the frequency of review during the project>
What is the status of mitigation plans?
<attach a copy of the high priority mitigation plans and their current progress>
What is the status of contingency plans?
<attach a copy of the high priority contingency plans and comment on which have been implemented>
What is the overall likelihood of achieving the project CSFs?
<comment on the chances of achieving the critical path of success>

(Continued)

Appendix 9.3 – The 'In Control?' Checklist (Continued)

Project Management Toolkit – 'In Control?' Checklist

Project:	<insert project title>	Project Manager:	<insert name>
Date:	<insert date>	Page:	2 of 2

Project organization

Are project activities being completed by the appropriate members of the organization?
<insert comment on the Project Team performance in terms of efficiency, effectiveness, capability and capacity versus the defined roles and responsibilities>

Contract and supplier management

What external suppliers are being used?
<comment on the status of the contract plan>
What is external supplier status and performance?
<confirm that procedures to manage supplier selection and performance are being followed, comment on supplier performance>

Project controls strategy

Are project costs under control? (Review cost plan – for example actual versus budget)
<comment on the current status of the cost plan, how regularly is this being reviewed, analyzed and reported?>
What is the likelihood that the project budget will be maintained (forecast to completion?)
<insert a current forecast of the final project cost versus the authorized budget>
Is the project schedule under control? (For example, review schedule and milestone progress)
<comment on the current status of the schedule, how regularly is this being reviewed, analyzed and reported?>
What is the likelihood that the project schedule will be achieved (forecast to completion?)
<insert a current forecast of the final project completion date versus the authorized schedule>
Are there any changes to scope (quantity, quality and functionality)? Are the costs and schedule under control?
<comment on the status of the change control process, attach a copy of the current change register>

Project review strategy

Are regular Stage Three reviews being conducted? (Is performance being managed and monitored?)
<comment on the schedule of reviews which have been planned and conducted, attach the previous review summary>
Is project performance adequate for project success?
<comment on project performance based on previous project review data plus this review>
Is there regular reporting? (Is the Project Team adequately managing communication of progress and performance to all stakeholders?)
<comment on the adequacy of progress reporting and communication to all internal and external stakeholders>

Stage Three decision

Is the project under control? (Is the project control strategy robust enough for project delivery to continue?)
<comment on the success of the control strategy in use>
What is the certainty that the project will be successful?
<comment on probability that this project will achieve its goals>

Appendix 9.4 – The 'Benefits Realized?' Checklist

Project Management Toolkit – 'Benefits Realized?' Checklist

Project:	<insert project title>	**Project Manager:**	<insert name>
Date:	<insert date>	**Page:**	1 of 1

Stage Three check

Have there been any changes since Stage Three completion? (Note only the changes since the final Stage Three 'health check')
<insert comments regarding any changes to the project since the previous health check>

Business benefits

Has the business case changed since Stage One (For example, during planning and delivery, pre- or post-project approval)
<review the formal business case and confirm its validity – insert any additional comments if there have been changes since approval – note if formal change agreements were made with the sponsor>
Have all benefits been defined in terms of trackable metrics? (Why is the project being done?)
<insert comments with regard to the status of the Benefits Specification Table and whether this has been converted into a Benefits Tracking Table>
What is the customer feedback? (Feedback from all stakeholders in the business including the customer)
<insert comments regarding any feedback from the customer particularly relating to the benefits delivery>
Are the benefits being tracked?
<attach a copy of the Benefits Tracking Table – comment on how long the benefits have been tracked>

Business change

Is the business ready for the project? (If the project can only enable benefits delivery by changing the way people work – has this been delivered, for example, training?)
<consider how the business has reacted to the changes delivered by the project – attach any completed sustainability reviews>

Scope definition

Has the scope been delivered?
<insert comments on the delivery of the project as agreed – refer to the CSFs>
Have the benefit enablers been delivered? (Are you sure that the project will enable the benefits to be delivered now the project is complete?)
<consider if the delivered project enables the benefits to be delivered as agreed by the business case>

Stage Four decision

Has the project been delivered? (Delivery of project critical success criteria)
<insert the decision – yes or no – with back-up material such as a completed after action review>
Have the business benefits been delivered? (Why was the project done in the first place?)
<insert the decision – yes or no – with back-up material such as a completed benefits scorecard>

Appendix 9.5 – The Project Management Toolkit

The concepts and tools introduced in *Project Management Toolkit* (Melton, 2007) are summarized in Table 9-1.

Table 9-1 The Project Management Toolkit concepts and tools

The Project Management Toolkit		
Project stage	**Concepts**	**Tools**
Stage One	➤ *Why* are we doing this project?	➤ 'Why?' Checklist ➤ Benefits Hierarchy ➤ Benefits Specification Table ➤ Business Case Tool
Stage Two	➤ *How* are we going to deliver the *what* of this project?	➤ 'How?' Checklist ➤ Table of Critical Success Factors ➤ RACI Chart ➤ Stakeholder Management Plan ➤ Control Specification Table
Stage Three	➤ Are we delivering this project *in control*?	➤ 'In Control?' Checklist ➤ Risk Table and Matrix ➤ Earned Value Tool ➤ Project Scorecard
Stage Four	➤ Have we delivered the *benefits*?	➤ 'Benefits Realized?' Checklist ➤ Benefits Tracking Tool ➤ Project Assessment Tool ➤ Sustainability Checklist

Appendix 9.6 – The Planning Toolkit

The concepts and tools introduced in *Real Project Planning* (Melton, 2008) are summarized in Table 9-2.

Table 9-2 The Planning Toolkit concepts and tools

The Planning Toolkit			
	Business planning (Chapter 2)	**Set-up planning (Chapter 3)**	**Control planning (Chapter 4)**
Concepts	⇒ Business strategy ⇒ Sponsorship – selection and role ⇒ Customer management ⇒ Consultancy process ⇒ Benefits management ⇒ Sustainability planning ⇒ Business change process and plan ⇒ Stakeholder analysis and planning ⇒ Communications strategy ⇒ Benefits mapping and specification ⇒ Environment assessment and change readiness	⇒ Set-up strategy ⇒ Project Manager and team selection ⇒ Project management capability model and capability profiles ⇒ Value definition and management ⇒ Scope quality, quantity and functionality definition and planning ⇒ Project and programme organization ⇒ Funding strategy ⇒ Team skills matrix and organization structures ⇒ Team start-up (team building and team processes) and performance planning ⇒ Project roadmaps and the stage gate approach ⇒ Critical success factors (CSFs) and critical to quality criteria (CTQ) ⇒ Scope risk assessment ⇒ Work breakdown structure and activity mapping ⇒ Business satisfaction analysis	⇒ Control strategy ⇒ Forecasting ⇒ Quality control ⇒ Contract and supplier strategy and planning ⇒ Risk management process and tools ⇒ Cost estimation, planning and contingency ⇒ Facilitation process and modes ⇒ Cost and schedule risk assessment ⇒ Schedule estimation, planning and contingency ⇒ Project progress measurement and performance management ⇒ Critical path of risks ⇒ Failure mode effect analysis (FMEA), risk flowcharts and checklists
Tools	⇒ Stakeholder Contracting Checklist ⇒ Sponsor Contract Planning Tool ⇒ Benefits Realization Plan ⇒ Sustainability Plan ⇒ Project Charter ⇒ Communications Planning Tool	⇒ Project Manager Selection Checklist ⇒ Team Selection Matrix ⇒ Project Team Role Profile ⇒ Team Start-up Checklist ⇒ Roadmap Decision Matrix ⇒ Finance Strategy Checklist ⇒ Activity Plan ⇒ Scope Definition Checklist	⇒ Project Scenario Tool ⇒ SWOT Table ⇒ Critical Path of Risks Table ⇒ Risk Management Strategy Checklist ⇒ Contract Plan ⇒ Supplier Selection Matrix

Appendix 9.7 – The Benefits Management Toolkit

The concepts and tools introduced in *Project Benefits Management* (Melton et al., 2008) are summarized in Table 9-3.

Table 9-3 The Benefits Management Toolkit concepts and tools

The Benefits Management Toolkit			
	Benefits concept **(Chapter 3)**	**Benefits specification** **(Chapter 4 and 5)**	**Benefits realization** **(Chapter 6)**
Concepts	<ul><li>Development of benefits criteria</li><li>Benefits, issues and activity mapping</li><li>Benefits measurement and scoring</li><li>Stakeholder management</li></ul>	<ul><li>Scope definition</li><li>Hierarchy of objectives</li><li>CTQ</li><li>Resistance to change</li><li>Business case development cycle</li><li>Lean thinking and lean value management</li><li>Cost/benefit analysis</li></ul>	<ul><li>Delivery of explicit and implicit benefits</li><li>Benefits risk assessment</li><li>Benefits tracking and cumulative scoring</li><li>Customer contracts and Kano analysis</li><li>Disengagement and project closure</li><li>Sustainability</li></ul>
Tools	<ul><li>Benefits Mapping Tool</li><li>Benefits Matrix</li><li>Benefits Scoring Tool</li></ul>	<ul><li>CTQ Scope Definition Tool</li><li>Scope Challenge Checklist</li><li>Business Environment Checklist</li><li>Benefits Influence Matrix</li><li>Business Case Template</li></ul>	<ul><li>Benefits Realization Risk Tool</li><li>Customer Satisfaction Analysis Tool</li><li>In-Place – In-Use Analysis Tool</li></ul>

Appendix 9.8 – Glossary

Benefits Criteria
The reason the project is being done; the articulation of benefits as linked to an organizational objective.

Benefits Enabler
What the project has to deliver to enable the benefits to be delivered; the project scope.

Benefits Hierarchy
A tool which confirms the alignment of the intended project scope to the targeted business benefits.

Benefit Metrics
Those measures which will confirm, after project delivery, that the business is realizing the benefits.

Benefits Mapping
A method which identifies and articulates the benefits that relate to the specific organizational goal.

CSF – Critical Success Factors
A quantifiable/measurable and identifiable action/activity that has the potential to impact the overall success of the project.

Critical Path of Success
The linkage of high level critical success factors to form a critical path of quantifiable/measurable actions/activities that have the potential to impact the overall success of the project.

DMAIC
Define, Measure, Analyze, Improve, Control. The DMAIC framework is an industry best practice business improvement methodology which is at the heart of Six Sigma, a process to evaluate the performance of a system and develop sustained improvement.

FMEA
Failure Mode and Effects Analysis. A structured method for identifying and analyzing failure modes/defects within a process, system or product.

JDI
Just Do IT. A term from process improvement activities indicating those activities that should just be carried out immediately to realise benefits.

Kaizen
Japanese for continuous improvement. Typically referenced in the context of a "blitz" approach where there is a focussed activity on a particular process or activity to identify and bring about a change.

Kanban
A visual signal for a process to start or to stop.

Lean Six Sigma
Lean Six Sigma is a business improvement methodology which combines tools and methodologies from both Lean (waste elimination) and Six Sigma (variation elimination).

(Continued)

Appendix 9.8 – Glossary (Continued)

Portfolio

A collection of projects using a common resource pool. These resources could be assets, people or funding, for example. There is a dependency between project resources which need to be used optimally and the portfolio is managed as a whole.

Programme

A set of interdependent projects working together to achieve a defined organizational goal. There is dependency between project outputs/benefits and the programme is managed as a whole.

Project

A bounded piece of work which is non-routine for the organization. It is not a part of business as usual (BAU) but has a defined start and end point (when it is integrated into BAU).

Theory of Constraints

Theory of Constraints often referred to as TOC (Goldratt, 1997) is a management methodology based on the identification of an organizational goal, the constraints to that goal (bottlenecks) and how the constraints need to be managed to achieve the goal.

Visual Factory

A method of displaying visually the current status of a process or activity so that everyone can understand the performance and take any actions as a result.

Appendix 9.9 – References

Adams, M., Kiemele, M., Pollack, L. and Quan, T. Lean Six Sigma: A Tools Guide, 2nd Edition, Air Academy Associates, Colorado Springs, USA (2004).

Goldratt, E.M. Critical Chain, The North River Press, Great Barrington, MA, USA (1997).

Kaplan, R.S. and Norton, D.P. The Balanced Scorecard, HBS Press, Boston, MA, USA (1996)

Kiemele, M.J., Schmidt, S.R. and Berdine, R.J. Basic Statistics: Tools for Continuous Improvement, 4th Edition, Air Academy Press, Colorado Springs, Colorado (2000).

Melton T. *Project Management Toolkit: The Basics for Project Success*, 2nd Edition. Elsevier, Oxford, Great Britain (2007).

Melton T. *Real Project Planning: Developing a Project Delivery Strategy*. Elsevier, Oxford, Great Britain (2008).

Melton T., Iles-Smith P. and Yates J. *Project Benefits Management: Linking Projects to the Business*. Elsevier, Oxford, Great Britain (2008).

Index